The
Female Brain

Second Edition

Conceptual Advances in Brain Research

A series of books focusing on brain dynamics and information processing systems of the brain.

Edited by **Robert Miller**, *Otago Centre for Theoretical Studies in Psychiatry and Neuroscience, New Zealand (Editor-in-Chief);* **Günther Palm**, *University of Ulm, Germany; and* **Gordon Shaw**, *University of California at Irvine, USA.*

Brain Dynamics and the Striatal Complex
  edited by *R. Miller and J.R. Wickens*

Complex Brain Functions: Conceptual Advances in Russian Neuroscience
  edited by *R. Miller, A.M. Ivanitsky and P.M. Balaban*

Time and the Brain
  edited by *R. Miller*

Sex Differences in Lateralization in the Animal Brain
  by *V.L. Bianki and E.B. Filippova*

Cortical Areas: Unity and Diversity
  edited by *A. Schüz and R. Miller*

Memory and Brain Dynamics: Oscillations Integrating Attention, Perception, Learning, and Memory
  by *E. Basar*

A Theory of the Basal Ganglia and Their Disorders
  by *R. Miller*

The Female Brain, Second Edition
  by *C.L. Darlington*

The Female Brain

Second Edition

Cynthia Darlington

CRC Press
Taylor & Francis Group
Boca Raton   London   New York

CRC Press is an imprint of the
Taylor & Francis Group, an **informa** business

CRC Press
Taylor & Francis Group
6000 Broken Sound Parkway NW, Suite 300
Boca Raton, FL 33487-2742

First issued in paperback 2017

© 2009 by Taylor & Francis Group, LLC
CRC Press is an imprint of Taylor & Francis Group, an Informa business

No claim to original U.S. Government works

ISBN 13: 978-1-138-11767-9 (pbk)
ISBN 13: 978-1-4200-7744-5 (hbk)

Library of Congress Cataloging-in-Publication Data

Darlington, Cynthia L.
    The female brain / Cynthia Darlington. -- 2nd ed.
        p. ; cm. -- (Conceptual advances in brain research)
    Includes bibliographical references and index.
    ISBN 978-1-4200-7744-5 (pbk. : alk. paper)
        1. Brain. 2. Women--Physiology. 3. Sex differences. 4. Neuroendocrinology. I.
Title. II. Series: Conceptual advances in brain research (unnumbered)
        [DNLM: 1. Brain--physiology. 2. Hormones--pharmacokinetics. 3.
Psychophysiology. 4. Sex Characteristics. 5. Sex Factors. 6. Women--psychology.
WL 300 D221f 2009]

QP376.D37 2009
612.8'2--dc22   2009009973

Visit the Taylor & Francis Web site at
http://www.taylorandfrancis.com

and the CRC Press Web site at
http://www.crcpress.com

For Tim.

Everyone needs a hero. You are mine.

Contents

Series Preface

The workings of the brain, including the human brain, are a source of endless fascination. In the last generation, experimental approaches to brain research have expanded massively, partly as a result of the development of powerful new techniques. However, the development of concepts that integrate and make sense of the wealth of available empirical data has lagged far behind the experimental investigation of the brain. The series of books entitled *Conceptual Advances in Brain Research (CABR)* is intended to provide a forum in which new and interesting conceptual advances can be presented to a wide readership in a coherent and lucid way.

The series will encompass all aspects of the sciences of brain and behavior, including anatomy, physiology, biochemistry, and pharmacology, together with psychological approaches to defining the function of the intact brain. In particular, the series will emphasize modern attempts to forge links between the biological and the psychological levels of describing brain function. It will explore new cybernetic interpretations of the structure of nervous tissue; and it will consider the dynamics of brain activity, integrated across wide areas of the brain and involving vast numbers of nerve cells. These are all subjects that are expanding rapidly at present. Subjects relating to the human nervous system as well as clinical topics related to neurological or psychiatric illnesses will also make important contributions to the series.

These volumes will be aimed at a wide readership within the neurosciences. However, brain research impinges on many other areas of knowledge. Therefore, some volumes may appeal to a readership, extending beyond the neurosciences. Books suitable for the series are monographs, edited multiauthor collections, or books derived from conferences, provided they have a clear underlying conceptual theme. To make these books widely accessible within the neurosciences and beyond, the style will emphasize broad scholarship comprehensible by readers in many fields, rather than descriptions in which technical detail of a particular speciality is dominant.

The next decades promise to provide major new revelations about brain function, with far-reaching impact on the way we view ourselves.

These great breakthroughs will require a broad interchange of ideas across many fields. We hope that the *CABR* series plays a significant part in the exploration of this important frontier of knowledge.

Preface

In September 2007, when I received the invitation to write the second edition of *The Female Brain*, I was thrilled. I was also a bit dubious . . . would there be enough new information to warrant a second edition? I ran some literature searches and to my delight discovered that, indeed, research on the female brain has progressed. I began to plan the second edition. One of the first decisions I had to make was how to treat the new literature in the area. It was clear from the first literature search that some of the new experimental results that had been published were really confirming the results of earlier studies cited in the first edition. I had a choice of trading a study that I had cited originally for a newer one saying the same thing, listing every single confirmation of the original result, or sticking with the original and adding new references only when they amplified or clarified the results of the earlier study. I decided to go with the latter, which seemed to be the fairest solution. As a researcher myself, I know how frustrating it can be when you publish a result, someone else confirms it, and they get the citations rather than your original experiment. As for listing every replication, I think readability would have suffered. I also decided to add new sections on pregnancy, stress, and sleep.

I knew from the beginning that this was going to be a very different writing experience in comparison to the preparation of the first edition. We had study-leave plans and conference travel already finalized. Whether it is apparent or not, there should be a very international flavor to this book. Bits of it were written in Japan, Germany, Australia, and the United States. A lot of it was written in France. In fact, my first "thank you" is to the staff of the Hotel du Danube in Paris. For two wonderful months they made me feel comfortable and welcome, and provided a lovely, elegant environment for writing. The whole experience was a dream come true.

Many people have offered support and encouragement along the way, and I am grateful to each and every one of them. My friends on the Multi-Region Ethics Committee have punctuated our meetings with words of encouragement. My colleagues in our research group have been so thoughtful and supportive that I don't even know how to formulate an appropriate thank you.

I must express my appreciation to Associate Professor Robert Miller, who is the editor of this series. For almost twenty years now, Robert has been a source of inspiration to me. Finally, thank you to Paul and Max. My husband, Paul, who is ceaselessly amazed at the peculiarities of my spelling and grammar, has proofread, comforted, dealt with builders, cooked, and kept me up to date on the rugby scores. Our cat, Max, provided input and distractions for the first edition. He's a little older and wiser now and prefers to nap on a quilt or snooze on my lap while I work.

To all of you, thank you so very, very much.

CD
Dunedin, NZ

The Author

Cynthia Darlington received her PhD in neurophysiology from the University of Sydney in 1987 and completed postdoctoral studies in neurophysiology at the University of Otago. She is now Head of Department, Department of Pharmacology and Toxicology, University of Otago, Dunedin, New Zealand. Her research interests focus on CNS changes associated with inner ear damage, including cognitive changes and memory dysfunction. She is deputy chair of New Zealand's Multi-Region Ethics Committee and is an active supporter of the New Zealand Neurological Foundation.

chapter 1

Introduction

The male and his muse

For centuries the beauty of the female form has been celebrated by painters, sculptors, songwriters, and poets. The artist has *his* muse, a real or imaginary woman, who provides inspiration to the creative genius within. Why this celebration of the female form, why a female muse for a male artist? Obviously, because the female body is different from the body of the male. But is this difference so obvious to everyone? Perhaps not.

We have arrived at the beginning of the twenty-first century, congratulating ourselves upon one hundred years of scientific achievement but entertaining a rather indeterminate view of female and male physiology. Overt sexual differences aside, the human body is often viewed as almost androgynous. Ironically, the organ with the greatest reason to differ between the sexes, the brain, is viewed as the most androgynous of all.

The assumption of equivalence between the female and the male brain is even apparent in much of the most recent scientific literature on brain function. Almost by convention, male animals are used in laboratory experiments in neuroscience, but the results of the research are considered applicable to both males and females. Whether it is an anatomical study of the structure of the brain, or a neurophysiological study on memory mechanisms, male animals are generally used. The reason given is often that females tend to be inconsistent in their responses as a result of the estrous cycle. Even in clinical drug trials in humans, females are often excluded from the early phases of testing. The reason? The risk of pregnancy, and that females tend to be inconsistent in their responses due to the menstrual cycle. The flaw in the reasoning is enormous: these very results will be applied to females, and yet it is only recently that concern has been expressed by some researchers that research on male brain function may not be so generally applicable to females.

Issues of gender have clouded the issue of basic biological differences between the sexes, in terms of brain function. The traditional girl-boy roles engendered in most societies often make it difficult to distinguish biology from culture. The argument is often made that girls have been socialized to perform well on some tasks but not on others. This is true, but it is does not address the basic question of functional differences.

Differences in brain function may arise as a result of different brain structure (in computer terms, different hardware). The differences may also arise from virtually the same brain structures performing in different ways (same computer but running different software). Or, to make things more complicated, and more interesting, slightly different hardware with slightly different software may be the answer. The purpose of this book is to address the question of structural and functional differences between the female brain and the male brain. Are there differences? How good is the evidence? Where do the differences lie? Are there differences in the neuroanatomy of females, and if so, where? Do females and males process information differently, and if so, how?

This book is divided into nine chapters. In this introductory chapter, Chapter 1, some of the background information necessary for understanding the contents of the remaining chapters will be introduced. This chapter is meant to serve as a guide, and a reference point, for the remainder of the book. For readers who already have a background in the physiological sciences, some sections will be redundant. Just skip them. For readers who do not have such a background, Chapter 1 should be something you can refer back to as needed.

Chapter 2 is about the history of the study of the female brain. Needless to say, it is short. It is really intended to put the current lack of knowledge into some kind of historical perspective; to try to provide an answer to the inevitable question, "How could this kind of intellectual neglect come about?"

Chapters 3 through 8 explore specific aspects of brain structure and function. These are the chapters that review the empirical evidence relevant to the different aspects of brain structure and function. By necessity, these chapters contain scientific detail that may be difficult for some readers. It is hoped that the explanations provided with the material in the chapters and in Chapter 1 will make the reading easier. Chapter 3 looks at the evidence for structural differences between the female brain and the male brain. Chapter 4 examines the evidence for functional differences in terms of neurotransmitters and their receptors. In many cases it is difficult, if not impossible, to separate structure from function, so the division between the chapters is somewhat arbitrary. Chapter 5 is about laterality, the functional asymmetry of the brain, the left brain–right brain distinctions and how they differ between females and males. In Chapter 6, the evidence for differences in perception and cognition are examined. How and why males and females apparently process memories differently is discussed. In Chapter 7, female/male differences in neurological and psychiatric disorders are discussed. In Chapter 8, we examine the treatment strategies for those disorders and the role that hormones may play. For Chapters 4 through 8, one of the important factors always under consideration is the menstrual cycle: how abilities and symptoms change with fluctuating hormones; and how treatment strategies may need to consider cyclic changes. The effects of hormone replacement therapy are also discussed.

Chapter 9 contains the summing up and conclusions. In the first edition, this was very much a "Where do we go from here?" chapter. Now, five years later, the situation has changed enough to make it possible to offer a more substantial direction for future research and also a preliminary "user's guide" for the female brain.

In case there is any misunderstanding, "different" is not a value judgment. It does not mean "better" or "worse." To the majority of readers, this will be self-evident and barely worthy of mention. I do mention it, however, because there seems to be a small group of people who think that to acknowledge difference is to acknowledge weakness. Surprisingly, many of the people holding this view are females. I don't know if this is something they really believe or something they really fear, but it is a response I find difficult to understand. Difference is the stuff of poetry, dreams, great art, and, above all, science. It is the essence of humanness.

The background

There are a number of different kinds of information that may be helpful to different readers. Some of the information is in the form of background information for different chapters. Other information is in the form of general guidelines, "rules for reading." The background information in this chapter includes the anatomy of the brain, neurotransmitters and their receptors, hormones, and the menstrual cycle. The "rules" include the use of terminology, abbreviations (including a list of common abbreviations), and referencing (what material is referenced to specific literature and what is not).

Hormones

The place to start the background discussions has to be with hormones. The endocrine system, which regulates and is made up of hormones, is one of the most complicated systems of the body. In order to keep the discussion of hormones even moderately brief, and to keep the length of this book to one volume, it is necessary to place strict limitations on the hormones and actions to be discussed. This necessary limitation, unfortunately, results in a hormone "shopping list." If you read the following section bearing in mind that the purpose of these hormones is to propagate the species, it may be a little easier to follow. For further information, a textbook of endocrinology or *Essential Reproduction* by Johnson and Everitt (1995) is recommended.

The major hormonal players that we are going to consider consist of estrogen and progesterone for females and testosterone for males. Interestingly, estrogen can be produced from testosterone, so although it is considered a "female" hormone, it is also of importance to males. These three hormones are synthesized from cholesterol via several different pathways (Figure 1.1). There are three forms of estrogen: 17β-estradiol, estrone, and estriol. Progesterone is a progestogen, as are 17α-hydroxyprogesterone and 20α-hydroxyprogesterone.

The release of estrogen, progesterone, and testosterone is controlled by the hypothalamus (see next section), via the secretion of gonadotrophic-releasing hormone. This hormone acts on the anterior pituitary gland to stimulate the release of follicle-stimulating hormone and leutinizing hormone, two other essential reproductive hormones for both females and males. In males, these two hormones regulate spermatogenesis and the production of testosterone. In females, follicle-stimulating hormone and estrogen control the development of the follicle and leutinizing hormone

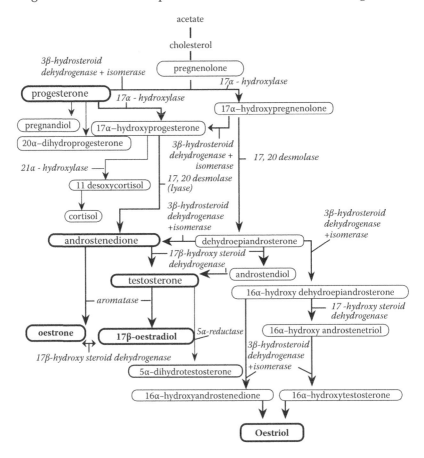

Figure 1.1 The synthesis pathways for the production of estrogen, progesterone, and testosterone. Adapted from Johnson and Everitt (1995).

regulates the secretion of estrogen. A peak in the release of leutinizing hormone triggers ovulation. Estrogen is synthesized from andros-
tenedione by the follicle of a mature oocyte. Hence, when the supply of
oocytes is exhausted, menopause occurs and the supply of estrogen ceases.
Progesterone is produced by the corpus luteum, a structure formed by the
ruptured follicle after the oocyte has been released. The corpus luteum
also releases estrogen; however, after about 12 days, it begins to deterio-
rate (regress) and the secretion of progesterone and estrogen decline.

The levels of estrogen and progesterone in the circulating blood fluctuate across the menstrual cycle (Figure 1.2) in the human female. Similar
fluctuations can be observed during the estrous cycle in female rats. In
terms of the menstrual cycle, estrogen rises during the early part of the
cycle (released by the mature follicle) and peaks just before ovulation.
After ovulation the level begins to drop, but there is another, lower, peak
before it drops sharply toward the end of the cycle. During the time leading up to ovulation, estrogen acts on the endometrium (the lining of the
uterus), causing it to thicken and become enriched with blood vessels in
preparation for implantation of a fertilized ovum. After ovulation, progesterone, released by the corpus luteum, begins to rise. If fertilization
does not occur, progesterone causes the enriched, but not required, endometrium to shrink and detach from the uterus. Toward the end of the
cycle, and by the start of menstruation, the levels of both estrogen and
progesterone are low (Walker 1997).

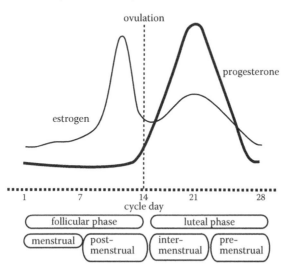

Figure 1.2 Schematic representation of the menstrual cycle showing relative levels of estrogen and progesterone. The phases of the menstrual cycle are shown
below the cycle representation.

If pregnancy occurs, instead of estrogen and progesterone dropping in the second half of the cycle, both hormones rise and continue to do so throughout the pregnancy. As full term approaches, the levels of estrogen and progesterone have increased from their normal cycle peaks by 300 percent and 1000 percent, respectively. Prolactin also increases throughout pregnancy, drops partially at delivery, and then is maintained at an elevated level throughout lactation (Figure 1.3). Lactation may delay the return to normal cycling, and estrogen levels will remain at baseline until cycling resumes. This does not, however, guarantee that fertility has not returned. It is not uncommon for a second pregnancy to occur within weeks of the first delivery, resulting in the birth of two children less than a year apart. One of the difficulties in understanding neuroanatomical and neurophysiological changes during pregnancy is that the methods to discover these changes are not amenable to use in human pregnancy. Scanning studies use radioactivity and contrast media and drugs, not substances that should be administered during pregnancy. Until a nonradioactive scan, devoid of contrast medium and drugs, can be developed, the best we can do is make educated guesses about the pregnant brain based upon our knowledge of nonpregnant hormone actions and cognitive measures.

The first concern with any drug administered during pregnancy is whether or not the drug is teratogenic — that is, whether or not it has the potential to cause birth defects. The FDA Classification of Teratogenicity has five classifications from A (no known teratogenic activity) to X (known to cause serious birth defects, considered too dangerous to administer during pregnancy). Drugs in the latter category include vitamin A analogues and thalidomide. A number of psychoactive drugs fall into category D (there is positive evidence of human risk but they may be necessary for

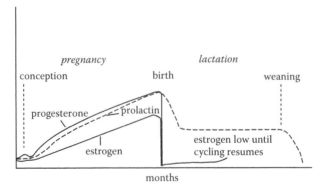

Figure 1.3 Schematic representation of hormonal changes during pregnancy and lactation. Hormone levels as shown do not represent actual physiological values.

the health and well-being of the mother). With category D drugs, it is a matter of weighing the pros and cons and making an informed decision on whether to administer the drug. This will be discussed further in the section on antiepileptic drugs. It is safe to say that all drugs can cross the placenta to some extent; sedatives administered to the mother will cause fetal sedation. In the worst case, drugs abused during pregnancy may lead to dependence in the fetus and a drug-withdrawal syndrome in the newborn baby.

Drugs taken during lactation also cross into the milk, so the baby may receive a dose of whatever the mother is taking. Prescribers are usually very careful to ensure that drugs given during lactation are safe for mother and child. For example, the antibiotic tetracycline is associated with discoloration of the developing teeth of the fetus when taken during pregnancy and may have a similar effect when the baby is exposed via breast milk.

It is the cyclic rise and fall of hormone levels that distinguishes female physiology. Males produce testosterone in a regular, pulsatile fashion. The male's blood level of testosterone peaks approximately every one and a half hours throughout the 24-hour period. This level of production remains about the same day in, day out, year after year, although there is some decline in testosterone production in old age. Circulating estrogen in males, which has been synthesized from testosterone, also remains at a fairly constant level. For females, from puberty to menopause, the hormonal milieu is constantly, though predictably, changing. So, for research in females, there are 28 (assuming a 28-day menstrual cycle) possible hormonal states to take into consideration. Ideally, all research in which the menstrual cycle is a variable would use blood analysis to determine hormone levels exactly. Due to practical constraints (such as finances), however, blood analysis is often not an option. Many researchers instead rely on their experimental subjects' keeping an accurate record of their menstrual cycles. The researchers then plan their experimental testing for particular phases of the cycle. The two most common divisions of the menstrual cycle are the two-phase cycle and the four-phase cycle (Figure 1.2). From Figure 1.2, it is easy to see that the two-phase cycle only provides information on what happens when progesterone is present or absent. The four-phase cycle allows for some distinction between the effects of estrogen and progesterone. Just to complicate things, the "normal" cycle length can be anywhere between 26 and 32 days. Ovulation does not always occur on day 14 (or even at all) and in some cycles, for some reason, progesterone does not rise normally. The female in question may be completely unaware that either variation from normal has occurred.

Another important characteristic of hormones, for experimental purposes, is that they are present in the blood circulation only in very small concentrations. Some, for example, are present in concentrations of approximately 10^{-9} molar, that is, there is one gram of the hormone in ten million liters of blood. Understandably, hormones at this concentration are very difficult to detect in the blood or other tissue by standard chemical analysis. The use of high-performance liquid chromatography has made the analysis of very low concentrations of hormones and neurotransmitters much more accurate; however, the problem of the constantly changing levels remains.

Some hormone researchers have a very direct solution to this problem. Animals used in hormone research have their gonads (ovaries in females, testes in males) removed prior to testing. The age at which the animals are gonadectomized will depend upon the questions being asked by the study. If, for example, the study is looking at receptor binding during infancy, then rat pups may be operated on right after birth. If, on the other hand, the study is intended to look at receptors in the fully developed adult, then the procedure may be performed a short time, for example, one week, before the binding study takes place. To study the "natural" presence of the receptors, a most unnatural condition is created. In females, the process of removing the ovaries is called, not surprisingly, ovariectomy, and female animals that have had the procedure are referred to as "OVX females" or just "OVX."

The central nervous system

After hormones, the next background section has to be about the brain. The following section gives a very basic description of the brain and spinal cord. For more detailed descriptions and information, *Principles of Neural Science* by Kandel et al. (2000), the "bible" of neuroscience, is highly recommended.

The central nervous system (CNS) is composed of the brain and spinal cord, not just the brain as many people assume. The spinal cord connects the brain to the peripheral nervous system and allows sensory information to be transmitted to the brain, and behavioral responses (motor commands) to be transmitted to the muscles and glands. Between sensory input and motor output, the brain analyzes the sensory signals for meaning, integrates this information with other sensory inputs and information stored in memory, and generates a motor command that causes the appropriate response to occur. The process of receiving sensory information and generating an appropriate response may be available for conscious analysis or it may be completely unconscious, depending upon the kind of information received and the type of response required. So, for example, we will consciously decide to take a piece of chocolate cake but will be completely unaware of our body's responses to the associated increase in blood sugar. Sometimes, however, we may have access to the results of the brain's capacity for unconscious processing. Most of us have had the experience of seeing a familiar face but not being able to remember the person's name. Then, sometime later, after the incident has been "forgotten," not only do we suddenly remember the person's name but often many other details about her as well. Such experiences have given rise to the well-worn joke, "I can't tell you right now, but I'll ring you at three in the morning when I remember."

In terms of gross anatomy, the brain can be divided into four regions: the *cerebral hemispheres* (including the cerebral cortex); the *diencephalon* ("the between brain"); the *brain stem*; and the *cerebellum* (Table 1.1) (Figure 1.4). These four main sections are connected by bundles of nerve fibers that

Table 1.1 The major divisions of the brain and some of their functions.

Division/subdivisions	Function
Cerebral Hemispheres	
Cortex	
Frontal lobe	Emotional, behavioral "control"
Temporal lobe	Language
Parietal lobe	Language, balance, hearing
Occipital lobe	Vision
Basal ganglia*:	Motor control, cognition
globus pallidus, putamen,	
caudate nucleus (globus pallidus + putamen = *lenticular nucleus*)	
(putamen + caudate nucleus = *striatum*)	
Limbic system*:	Memory and emotion
amygdala, cingulate,	
hippocampus, septum	
Diencephalon	
Thalamus	Sensory, motor processing and relay
Hypothalamus	Control of homeostasis
Brain stem	
Tegmentum	Autonomic control
Reticular formation	Arousal
Midbrain	
Periaqueductal gray	Pain perception
Superior colliculus	Eye-head coordination
Substantia nigra	Motor control
Inferior colliculus	Auditory reflexes
Pons and medulla	
Nuclei of cranial nerves	Sensory and motor function
Inferior olive	Cerebellar relay
Cerebellum	Fine motor control

*There is some disagreement about which nuclei should be included in these divisions.

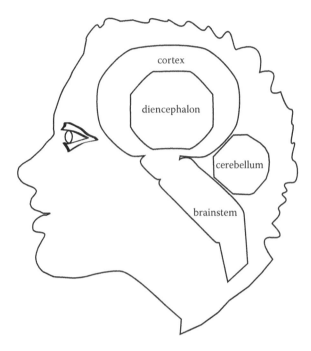

Figure 1.4 Schematic representation of the four divisions of the brain.

carry information to and from the different regions of the brain and to the spinal cord. These complex networks of interconnections allow the activities of the different brain regions to be coordinated smoothly and efficiently. For a comprehensive coverage of the structure of the human brain, Kandel et al. (2000) and Diamond et al. (1985) are recommended.

The cerebral cortex is the outermost area of the top and sides of the brain. Viewed from the outside, the cortex appears as a wrinkled and folded surface covering the rest of the brain. When the cortex is cut to expose a cross-section, thick layers of neatly ordered cells can be seen. These layers are indeed folded into deep grooves and wrinkles that cover the cortical surface. In phylogenetic terms, the cerebral cortex is the newest part of the brain. That is, it is the most highly evolved area and is the site of the highest levels of cognitive function. The folding of the cortical surface is an important aspect of brain evolution. The deep folds allow for greater surface area in each cortex and, therefore, a greater number of cells and greater capacity for information processing. The cortex of a mouse or rat, for example, has fewer and shallower folds than the cortex of a primate, such as a lemur, which in turn has fewer folds than the cortices of humans.

The hemispheres, including the cortex, are paired. The left and the right hemispheres have different functional specialties, often described as logic and creativity, respectively. The areas of each cortex are further distinguished anatomically by very deep folds, known as sulci, which loosely define the boundaries of four lobes: frontal, temporal, parietal, and occipital, named for the bones of the skull that cover them (Figure 1.5). Areas of the cortices are also organized functionally into primary, secondary, and tertiary regions. The primary regions are those areas most concerned with our interactions with the world. So the primary sensory areas are those that receive sensory information mainly from a single sensory system via the peripheral nervous system — for example, the visual system. The primary motor cortices are those areas that send the commands for action directly to the spinal cord to produce appropriate behavioral responses. The secondary and tertiary regions surround the primary regions and conduct complex processing that may include information on several aspects of the sensory signal being processed. One of the cortical regions best understood in terms of the increasing complexity of sensory information processing is the visual cortex. In 1965 Hubel and Wiesel won the Nobel Prize for their work in which they identified rows and columns of neurons in the visual cortices of cats, which processed increasingly complex aspects of a visual stimulus.

Finally, there are the association areas. In the association areas, all of the information necessary for the planning and initiation of movement

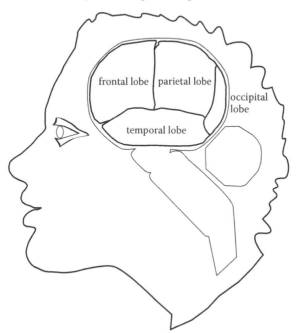

Figure 1.5 Schematic representation of the lobes of the cerebral cortex.

and other responses is integrated. Information from the sensory systems is combined with information retrieved from memory about previous encounters with the particular stimulus; previous responses; new information that may have changed the meaning of the stimulus since the last time it was encountered; and finally, the desirability of different response outcomes. It is from the association areas that the commands to initiate the chosen behaviors are sent to the tertiary motor areas and ultimately to the primary motor area where the movement command is issued.

The two other areas of the hemispheres of particular interest are the basal ganglia and hippocampus (Figure 1.6). The basal ganglia (and here there is some disagreement over what specific nuclei should be included) are important in movement control, and they also play a role in cognitive function. The limbic system (again, somewhat controversial in terms of what should be included), composed of the hippocampus and amygdala, cingulate, and septum, is said to be the "emotion-control center" of the brain. It plays a role in control of the autonomic nervous system, and indirectly modulates some hormone levels in the body. The hippocampus is also intimately involved in memory.

Underneath and between the cerebral hemispheres is the diencephalon. This area includes principally the thalamus and the hypothalamus. The primary function of the hypothalamus is to maintain homeostasis, a

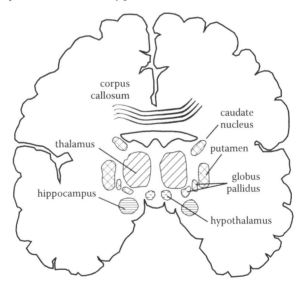

Figure 1.6 Schematic representation of a cortical section of the brain, approximately at the level of the junction of the frontal and parietal lobes. Cross-hatch represents the basal ganglia, horizontal lines represent the hippocampus, diagonal lines represent the thalamus, and broken lines represent the hypothalamus.

state of physiological balance, which includes optimal body temperature, blood pressure, heart rate, and respiration. The thalamus is the relay center for information being sent to the cerebral hemispheres. Some of the information relayed through the thalamus is almost unchanged from the original sensory signal, but other information undergoes processing in the thalamus before it is relayed to the cortex.

The hypothalamus is the major CNS control center for the endocrine system, the hormonal system of the body, which is essential for the control of homeostasis. The hypothalamus controls hormonal activity throughout the body by synthesizing and releasing hormones from secretory cells directly into the bloodstream and also by neuronal activity transmitted by neurons in the hypothalamus. This type of dual control allows the hypothalamus to have both slow, and long-lasting, control of hormone levels as well as rapid, neuronal responses. With these two distinct types of response, long-term changes, such as the menstrual cycle, and short-term changes, such as the "fight or flight response" to danger, can be accurately maintained.

The hormones released from the hypothalamus regulate the release of other hormones from the pituitary gland (Table 1.2). Through a number of complex feedback mechanisms, the hypothalamus closely regulates and monitors hormone secretion to ensure the correct hormonal balance throughout the body. In addition to the synthesis and release of hormones, the hypothalamus also controls or regulates a number of other functions. These include the balance of salt and water (electrolytes) in the body; the autonomic nervous system; appetites including thirst and sexual desire; body temperature and respiration; emotions; and the maintenance of a number of biological rhythms. Not surprisingly, it is in the hypothalamus that scientists first looked for differences in brain structure between females and males.

The cerebellum sits at the back of the brain, behind and below the cerebral hemispheres. Like the cortex, it is composed of precisely organized

Table 1.2 Hypothalamic regulation of the release of pituitary hormones.

Target	Hormone
Anterior pituitary	Growth hormone–releasing hormone
	Thyrotropin-releasing hormone
	Corticotrophin-releasing hormone
	Gonadotrophin-releasing hormone
	Growth hormone release–inhibiting hormone
	Prolactin release–inhibiting hormone
Posterior pituitary	Antidiuretic hormone
	Oxytocin

layers of neurons folded into deep ridges. The cerebellum is essential for almost all aspects of fine motor control, including eye movements, balance, and posture. All highly skilled movements, for example, tap dancing or playing the piano, are dependent upon the cerebellum for precision and accuracy.

After the cerebral hemispheres, the diencephalon, and the cerebellum, the brain stem is what is left over. It is an elongated structure that extends along the lower surface of the brain from just behind the diencephalon to under the cerebellum, where it continues without interruption to form the spinal cord. Phylogenetically, the brain stem is the oldest part of the brain, and this part of the brain is remarkably similar not only in mammals, but also in nonmammalian species. Loosely speaking, the rest of the brain has evolved on top of the brain stem. This is the area where the control centers for vital functions such as cardiovascular and respiratory control are found. The area closest to the diencephalon, the midbrain, is involved primarily in motor control. Behind the midbrain, the pons is also involved in motor control but, in addition, acts as a relay between the cerebellum and the cerebral hemispheres. It is through the pons that the cerebellar signals necessary for the fine-tuning of movements are relayed to the motor areas of the cortex. Behind the pons is the medulla. Together with the pons, the medulla contains the centers for the control of breathing and cardiovascular function, including the control of blood pressure. All three areas of the brain stem also contain a neuronal network, known as the reticular formation, which is important for the modulation of arousal, the overall state of awareness of an individual.

The spinal cord continues downward from the medulla and carries signals from the brain to the nerves of the peripheral nervous system. The cord is arranged into ascending tracts (dorsal columns), which carry information from the peripheral nervous system to the brain, and the descending tracts (ventral columns), which carry motor commands from the brain to the peripheral nervous system (Figure 1.7). Between these areas are groups of neurons that function within the spinal cord itself and mediate spinal reflexes such as the "knee-jerk" response to a tap of the patella tendon. The spinal cord is encased in the vertebrae of the spinal column, which act as a coat of armor, just as the skull protects the brain. This extensive form of protection is necessary because breaks in or damage to the spinal cord result in a loss of sensation and/or motor control below the level of the lesion.

The nerves supplying different parts of the body exit from the spinal cord to the peripheral nervous system at very specific positions so that the area of the body affected by damage to the spinal cord can be accurately predicted from the known location of a lesion. The nerves leave the spinal cord between the vertebrae, and vertebrae are numbered according to their position in the cervical, thoracic, or lumbar portion of the spine. So,

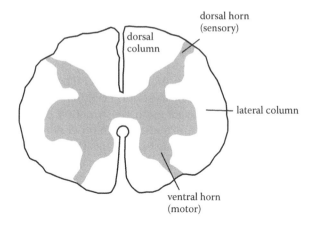

Figure 1.7 Schematic representation of a cross-section of the spinal cord. Shaded areas represent the dorsal and ventral horns.

for example, a complete lesion of the spinal cord at the level of thoracic vertebrae 3 (T3) will consistently result in a complete loss of sensation and motor control below the mid-chest level.

The entire brain and spinal cord are surrounded by three layers of membranes, the *meninges*, which form a protective sac. Within this protective sac, and also within cavities within the brain (ventricles), *cerebrospinal fluid* (CSF) circulates. CSF is essential to the function of the brain. In addition to providing a fluid cushion to prevent the brain from being damaged by knocking against the inside of the skull, the CSF bathes the brain in nutritive substances and also carries away waste products. In patients with suspected neurological diseases, analysis of the CSF is an important tool in the diagnostic process. Small quantities of CSF may be extracted by inserting a needle into a space at the bottom of the spine and withdrawing a small amount of fluid in a procedure known as a lumbar puncture. While not generally viewed as a patient's favorite life experience, the procedure can provide potentially life-saving information on the type of disease present. In the case of bacterial meningitis, for example, the strain of bacterium causing the infection can be identified and the appropriate antibiotic treatment initiated. Another protection for the brain is the *blood-brain barrier*. The blood-brain barrier refers to the arrangement of the cells in the walls of capillaries in the brain. The epithelial cells are positioned so that there is very little space between them ("tight junctions"), which prevents the movement of many substances from the blood into the brain. There are isolated areas of the brain, the *periventricular* organs, located on the margins of the ventricles, where the epithelial cells do not have tight junctions. Some substances can reach the brain via the periventricular organs.

Several different types of cells are found in the brain and spinal cord, but the principal type of cell is the *neuron*, the cell type with the capacity to transmit and receive information. It is the communication between neurons that constitutes the activity of the brain. Although neurons have many variations in shape and size, there is a general anatomical structure that allows a generic neuron to be described (Figure 1.8). A neuron may be divided into three areas based upon the function of each area. The dendrites are the areas where information from other neurons is received. The dendrites are often described as branching like trees, and they may spread quite widely to maintain contact with a large number of other neurons. The axons of other neurons may make synaptic contact with the dendrites and/or the cell body. The cell body (soma) is the information-processing center for the neuron. All of the different inputs arriving via the dendrites are processed and integrated to produce an appropriate response by the neuron. If the majority of the incoming signals received are excitatory, the neuron will respond by producing an electrical signal that will travel down the axon and cause the release of neurotransmitter at the synapse. If, on the other hand, the majority of signals received are

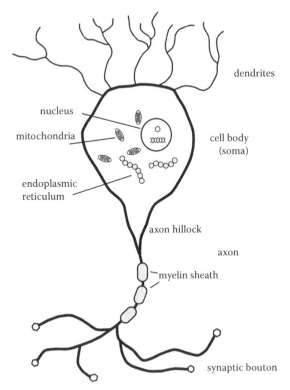

Figure 1.8 The generic neuron.

inhibitory, the neuron will not produce an electrical signal and the neurotransmitter will not be released.

The soma contains the internal organs of the neuron that maintain the neuron's health and function by manufacturing new proteins to replace worn parts, generating energy to support the neuron's activities, and by manufacturing and releasing neurotransmitters and/or neuromodulators. Extending from the cell body, usually on the side approximately opposite to the dendrites, is the axon. The axon carries the electrical signals (the action potentials) generated by the cell body to the synaptic terminals where the neuron makes contact with the dendrites or cell bodies of many other neurons. The neurons on which the terminals synapse may be within the same part of the brain or they may be located at a great distance from the signaling cell. Axons that must travel over large distances often travel together in bundles known as fiber tracts. One of the major fiber tracts is the corpus callosum, which conveys information back and forth between the two cortical hemispheres. When the corpus callosum is cut, one side of the brain literally does not know what the other side is doing. Some axons consist of a single fiber with few branches, but other types of axons may branch repeatedly, allowing access to a large number of neurons and even different brain regions. The axon endings consist of synaptic terminals that terminate close to the dendrites and in some cases the somata or axons of other neurons. The point of contact is known as the synapse, and the physical space between the synaptic terminal and the other neuron is known as the synaptic cleft. Neurotransmitters released from the synaptic terminal cross the synaptic cleft and bind to receptors on the dendrites or cell bodies that are located in regions known as synaptic densities (Figure 1.9). In this way signals are conveyed from one neuron to another.

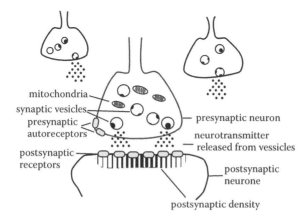

Figure 1.9 The generic synapse.

The majority of axons are wrapped in an "insulating" sheath called myelin that allows the signals to travel faster along the axon. In the CNS, myelin is formed by cells known as oligodendrocytes, which wrap themselves around the axons to form a sheath. Damage to the oligodendrocytes, as in multiple sclerosis, causes disruption of the myelin sheath and slows the signals traveling along the axon. When this occurs in a number of axons, communication between neurons may be severely disrupted.

Neurotransmitters are the chemicals released by the axon terminals of neurons. There are certain characteristics of neurotransmitters that define a "classical" neurotransmitter (Kandel et al. 2000). First, the chemical must exist in the presynaptic terminals. Second, the chemical must be released from the presynaptic terminal. Third, when the chemical is applied experimentally, it must have the same effect as the naturally occurring chemical. Fourth, there must be a mechanism for breaking down or removing the chemical from the synapse. Fifth, an *antagonist*, a substance that blocks the action of the naturally occurring chemical, must also block the action of the experimentally applied chemical. Thirty years ago a discussion of neurotransmitters was easy. There were a small number of known neurotransmitters that had a few different subtypes of their respective receptors. Now there are numerous substances that are known to act as neurotransmitters and for which there are a number of receptors and their subtypes (Table 1.3). In addition, it is now known that a number of substances that are not classical neurotransmitters (for example, hormones) can act as modulators of neurotransmitter systems.

Receptors are the sites where neurotransmitters released into the synaptic cleft bind, to convey information from the presynaptic neuron to the postsynaptic neuron. Receptors are specific structures, made of proteins, that span the cell membrane and convey information from outside the neuron to the intracellular space where the process of responding to the neurotransmitter begins. Different types of receptors function in different ways. However, there are some basic principles of receptor function (Figure 1.10). First, receptors do not respond to any neurotransmitter that happens to drift by. Their responses are specific to a particular type of neurotransmitter. This response specificity is termed "selectivity." Second, there is the principle of "saturability." There are a finite number of receptors, and once they are occupied, no more binding can take place. In receptor binding studies, it is important to establish saturability in order to demonstrate that what may seem to be binding is not just the drug or neurotransmitter being absorbed by other tissue. Third, receptor binding usually displays reversibility. This is necessary in order to act as an effective signaling system. The message is conveyed via neurotransmitter binding, then the neurotransmitter dissociates from the receptor. The release of the neurotransmitter serves two important functions: it

Table 1.3 Examples of some "classical" neurotransmitters and their receptors.

Neurotransmitter	Receptors	Actions
GABA	$GABA_A$	Increase Cl^- conductance
		Decrease cAMP
	$GABA_B$	Increase K^+ conductance
		Increase Ca^{2+} conductance
	$GABA_C$	Increase Cl^- conductance
Glycine		Increase Cl^- conductance
Glutamate	AMPA	Increase K^+ conductance
	$GLU_{(1-4)}$	Increase Na^+ conductance
	$GLU_{(5-7)}$	
	NMDA	
	$mGLU_{(1-7)}$	Decrease cAMP
		Increase IP_3/DG
Acetylcholine	Nicotinic	Increase K^+ conductance
		Increase Na^+ conductance
		Increase Ca^{2+} conductance
	Muscarinic	
	$mACh_{(1,3)}$	Increase IP_3/DG
	$mACh_{(2,4)}$	Decrease cAMP
		Increase K^+ conductance
Dopamine	$D_{(1,5)}$	Increase cAMP
	D_2	Decrease cAMP
		Increase K^+ conductance
		Decrease Ca^{2+} conductance
	D_3	?
	D_4	?
Noradrenaline	α_{1A-D}	Increase IP_3/DG
	α_{2A-C}	Decrease cAMP
		Increase K^+ conductance
		Decrease Ca^{2+} conductance
	β_{1-3}	Increase cAMP
Serotonin	$5\text{-}HT_{1A-F}$	Decrease cAMP
		Increase K^+ conductance
	$5\text{-}HT_{2A-C}$	Increase IP_3/DG
	$5\text{-}HT_3$	Increase K^+ conductance
		Increase Na^+ conductance
	$5\text{-}HT_4$	Increase cAMP
	$5\text{-}HT_{6\,and\,7}$	Increase cAMP

Source: Modified from Hardman et al. (1996). cAMP, cyclic adenosine monophosphate; Ca^{2+}, calcium; Cl^-, chloride; IP_3/DG, inositol triphosphate, diacylglycerol; K^+, potassium; Na^+, sodium.

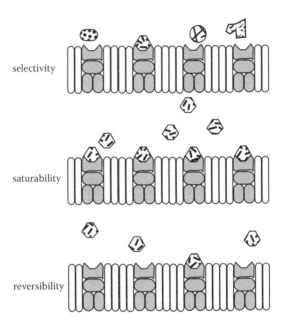

selectivity

saturability

reversibility

Figure 1.10 Schematic representation of the characteristics of receptor binding. Gray areas represent the receptor complex.

signals "message over" and it allows the neurotransmitter to be broken down and recycled.

However, and this is a rather big "however," the responses of specific types of receptors may be modified by substances other than their neurotransmitters, for example, drugs. The neurotransmitter GABA (gamma-aminobutyric acid), for example, causes an inhibitory response when it binds to a specific subtype of GABA receptor. When GABA binds to the "A" subtype of the GABA receptor (the $GABA_A$ receptor), it will cause a decrease in the activity of the cell on which the receptor is located. But the $GABA_A$ receptor, in addition to the binding site for GABA, also has a number of other binding sites that recognize different types of drugs, including alcohol, steroids, barbiturates, and benzodiazepines. When, in addition to GABA, one of these drugs is present (and bound to its binding site), the action of GABA on the receptor will be modified. Binding sites on receptors, which modify the actions of the neurotransmitter, are known as allosteric binding sites.

The presence or absence of allosteric binding sites on receptors is one way in which the action of a particular neurotransmitter may be made more specific. Another way specificity of action is achieved is by the distribution of the different receptor types, and subtypes, throughout the CNS. The number and types of neurotransmitter receptors that are present on neurons varies widely throughout the brain. Some neurotransmitter receptors are located only in very discrete brain regions, while other receptor types are widespread.

Interactions between hormones and neurotransmitters

It is now known that hormones and neurotransmitters interact at the receptor level (Chapter 4). One fascinating area where this interaction appears to be particularly important is in the development of the brain. This is also an area that is, by necessity, completely outside the scope of this book. For readers interested in the role that hormones and neurotransmitters play in the developing nervous system, there are two excellent reviews, one by Kawata (1995), the other by Döhler (1991), which are highly recommended.

Neuroscience methods

Many standard methods are used in neuroscience research and it is not possible to cover them all in this chapter. There are, however, two relatively new methods, based on recent advances in genetic research that may be unfamiliar to the reader.

The first method is RT-PCR. This is a biochemical method that is now widely used to examine the effects of various drugs or other variables on the activities within the neuronal nucleus that lead (under most conditions) to the expression of new protein. "RT-PCR" stands for "reverse transcriptase polymerase chain reaction." It is a high-resolution method for determining the extent to which a particular gene has been expressed. When a gene is activated, messenger RNA (mRNA) is produced from DNA in a process called "transcription." The mRNA template then travels to the ribosomes to produce the corresponding protein in a process known as "translation." In RT-PCR, the enzyme "reverse transcriptase" is used to make corresponding double-stranded copy DNA (cDNA). This is the reverse of the process that usually occurs during transcription. The PCR part of the technique relates to identifying and amplifying enough of a specific gene so that it can be quantified. Because DNA has a double-helical structure, it has the capacity to unwind and reform double-helices. The PCR method uses repeated temperature changes (thermal cycling) to do this. When DNA is heated to 95° C, the strands separate. Partial DNA sequences (*oligonucleotides*) for the gene of interest are synthesized, and these oligonucleotide primers bind or *hybridize* to complementary sequences in the single-stranded DNA. With large amounts of DNA polymerase, the partial DNA sequence of the primers will be "filled in" to complete the sequence. The process is then repeated thousands of times. Each time the separated DNA strands serve as a template for the construction of the DNA sequence or gene of interest. The gene expression can then be quantified using gel chromatography or by measuring the amount of a fluorescent probe inserted in the primer.

Genetic modification is another new technique that, with the rise of the biotechnology industry, is rapidly becoming accessible to many researchers (providing that they have the necessary funds to buy the materials). Mice may be genetically modified to express an extra copy of a gene or to express a new gene (*transgenic mice*). Conversely, mice may be genetically modified so that they do not express a particular gene at all (*knockout mice*). Production of genetically modified mice seems amazingly straightforward, but the technology involved would have dazzled mid-20th-century science fiction writers (and the success rate, at present, is only about 10 percent). A section of cDNA is produced that has been modified to either delete the sequence containing the gene being studied, or to include a new or additional sequence of a particular gene. The truly amazing part is that once the section of cDNA has been produced, it is simply injected into a fertilized mouse ovum. The ovum is then implanted into a surrogate mother along with unmodified ova from the same source. When the litter is born, it consists of both the experimental, genetically modified animals and their normal (wild-type, WT) littermates.

The rules

In writing a book of this type it is very difficult to decide upon, and adhere to, a particular set of rules for terminology and usage. Consider the use of the term "female." There are a number of ways to indicate that you are referring to the nonmale members of the human species. In addition to being "females," they can be "women," "ladies," "gals," "sheilas," or even "girls." When the aim is to maintain consistency throughout nine chapters, it is important to make the decision early and stick to it. In this case, "female" has been chosen and is used throughout the book. Sometimes, it sounds a little awkward and sometimes it sounds contrived. Overall, however, it works and at least it is accurate.

Also, for the sake of accuracy, it is tempting to spell out every single term and to distinguish between every possible form of a drug or chemical. For the sake of readability, it is necessary to decide upon an accurate, but palatable, usage of scientific names and terminology. Consider, for example, a discussion of the two hormones estrogen and progesterone. It is accurate to say, "The hormones 17β-estradiol, estrone, estriol, progesterone, 17α-hydroxyprogesterone, and 20α-hydroxyprogesterone have been demonstrated to fluctuate across the menstrual cycle of the human female but there is no estrous cycle in the ovariectomized female rat." It is much more comfortable (and for most purposes) just as informative to say, "The hormones estrogen and progesterone fluctuate across the menstrual cycle of the human female but there is no estrous cycle in the OVX rat."

In this book, when the important message is that an estrogen is involved in a particular process, the term "estrogen" will be used. If, however, the particular form of estrogen is important to the discussion, for example, in the case of a behavioral study comparing the actions of 17β-estradiol and estriol, then the type of estrogen will be specified. This same rule applies to the name "progesterone" and to certain classes of drugs, for example, benzodiazepines.

A similar rule has been applied to the referencing of scientific data. It is essential that experimental results and the ideas regarding their interpretation be referenced to the appropriate author. It is often the case, however, that several authors report similar results over a number of years. To reference all of the literature, for every chapter, would create a huge bibliography that would be of little interest or use to the majority of readers. Instead, I have adopted the following policy: when a particular result has been published by a number of authors, then very recent work that cites the previous publications is used as the reference. When it is the case that early work is relevant to a particular section, then that work will be cited, often in addition to a recent review. In the case of general material, for which a very wide literature exists, then a general reference source, usually a textbook, is cited. Finally, there are a number of areas that, while extremely interesting and important, are outside the scope of the book. For these, recent reviews have been recommended.

The results of the experiments discussed in this book will be, unless very clearly stated otherwise, based on the statistical analysis of the data, using the criteria for significance specified in the methods. Results that are "trends," "only marginally insignificant," or "encouragingly close to significant" will be reported as "not significant."

As far as possible, I have avoided using abbreviations. To read and understand a paragraph riddled with acronyms is difficult for a person who knows what they mean, but it is almost impossible for one who is not familiar with them. Sometimes, however, it is awkward and impractical not to use abbreviations. This is particularly true in the case of subtypes of neurotransmitter receptors. The choice is between spelling out the entire name, for example, "dopamine receptor subtype 3," every time it appears (which may be several times in a single sentence) and using the standard receptor shorthand, "D_3." The rule is this: when one or more letters appear followed by a subscript number and/or letter, it specifies a neurotransmitter receptor subtype. There are some standard abbreviations that appear routinely in the scientific literature and because of their common usage have virtually become terms in themselves, for example, "CSF" for cerebrospinal fluid. Although every abbreviation is identified the first time it is used, a table of commonly used abbreviations is provided as a reference, if needed (Table 1.4).

Table 1.4 Commonly used abbreviations.

Abbreviation	Meaning
ACh	acetylcholine
AMP	adenosine monophosphate
cAMP	cyclic adenosine monophosphate
BBB	blood-brain-barrier
CB (followed by subscript)	cannabinoid receptor
CNS	central nervous system
CSF	cerebrospinal fluid
CT	computerized tomography
D (followed by subscript)	dopamine receptor
DNA	deoxyribonucleic acid
E	estrogen
EEG	electroencephalography
GABA	gamma-amino-butyric acid (when used alone)
	gamma-amino-butyric acid receptor (when followed by subscript numbers and/or letters)
Glu (followed by subscript)	glutamate receptor
HRT	hormone replacement therapy
5-HT (followed by subscript)	serotonin receptor
MRI	magnetic resonance imaging (anatomical information)
fMRI	functional magnetic resonance imaging (functional information)
NA	noradrenaline
OVX	ovariectomy
mRNA	messenger ribonucleic acid
RT-PCR	reverse transcriptase polymerase chain reaction

Bibliography and recommended readings

Diamond, M. C., A. B. Scheibel, and L. M. Elson. 1985. *The Human Brain Coloring Book*. New York: Barnes and Noble.

Döhler, K. D. 1991. The pre- and postnatal influence of hormones and neurotransmitters on sexual differentiation of the mammalian hypothalamus. *International Review of Cytology* 131: 1–57.

Hardman, J. G., L. E. Limbird, P. B. Molinoff, R. W. Ruddon, and A. G. Gilman, eds. 1996. *Goodman & Gilman's The Pharmacological Basis of Therapeutics*. 9th ed. London: McGraw-Hill.

Johnson, M. H., and B. J. Everitt. 1995. *Essential Reproduction*. 4th ed. London: Blackwell Science.

Kandel, E. R., J. H. Schwartz, and T. M. Jessell. 2000. *Principles of Neural Science*. 4th ed. London: Prentice-Hall.

Kawata, M. 1995. Roles of steroid hormones and their receptors in structural organization in the nervous system. *Neuroscience Research* 24: 1–46.

Walker, A. E. 1997. *The Menstrual Cycle*. London: Routledge.

Wilson, J. D., and D. W. Foster, eds. 1992. *Williams Textbook of Endocrinology*. 8th ed. Philadelphia: W.B. Saunders.

chapter 2

A historical perspective

Taking a historical perspective on the study of the female brain is a little like pursuing the dog that didn't bark in the night. The real significance is what *did not* happen. To make the task of identifying causes even more difficult, it is not one but several things that did not occur. Certainly, the greatest omissions have occurred in the last hundred years, but the social perspectives that led to these omissions have developed over centuries.

In any society the areas chosen for research and development will reflect the needs, desires, and perspectives of the members of the society. Opportunity and resources will usually be available only for the development of priority areas of concern. Only in times of great wealth and excess can a society afford the luxury of the pursuit of knowledge just for the pleasure of knowing. The kinds of areas developed will depend in part upon the education and skills of the people doing the research and the personal interests and biases of those individuals. It follows that the present lack of knowledge of the female brain must stem from a number of circumstances where decisions were made that did not favor such interests. Since, until very recently, study of this type was almost exclusively the domain of physicians, the role of females in medical science will be an important aspect of the discussion.

Since ancient times females have been cast in the role of caregivers. (See Figure 2.1 for the historical progression of female education.) We know that the Neanderthals cared for their sick and elderly, but we do not know whether the caring role was exclusively for females. We do know that by the time females developed agriculture around 8000 BC and farming communities began to emerge, females were the farmers and caregivers while males were the hunters. It seems likely, at least in the early social orders, that females developed the role of healers.

At some stage, the ability to heal became endowed with mystical connotations and, not surprisingly, became a source of power. It was probably at that time that males began to take on the roles of healers. We do not know whether in the very early human societies males and females had equal social stature. We do know, however, that by around 1450 BC, the females in Mesopotamia had become the "other sex" and were under the control of males. A Mesopotamian female was obedient to her father, her husband, her father-in-law, and eventually, even to her sons.

The earliest female healers in recorded history were the female physicians of ancient Egypt (Brooke 1993). In 1900 BC the Kahun Papyrus,

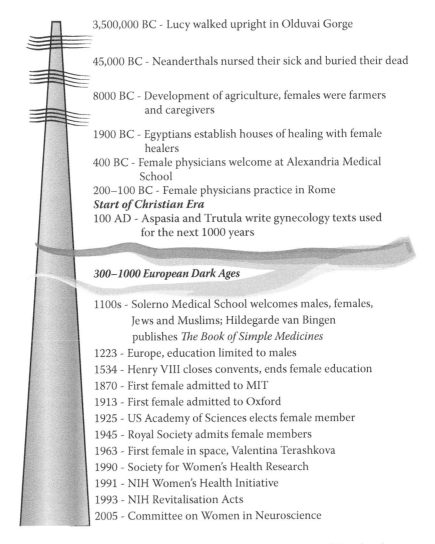

3,500,000 BC - Lucy walked upright in Olduvai Gorge

45,000 BC - Neanderthals nursed their sick and buried their dead

8000 BC - Development of agriculture, females were farmers
    and caregivers

1900 BC - Egyptians establish houses of healing with female
    healers

400 BC - Female physicians welcome at Alexandria Medical
    School

200–100 BC - Female physicians practice in Rome

Start of Christian Era

100 AD - Aspasia and Trutula write gynecology texts used
    for the next 1000 years

300–1000 European Dark Ages

1100s - Solerno Medical School welcomes males, females,
    Jews and Muslims; Hildegarde van Bingen
    publishes *The Book of Simple Medicines*

1223 - Europe, education limited to males

1534 - Henry VIII closes convents, ends female education

1870 - First female admitted to MIT

1913 - First female admitted to Oxford

1925 - US Academy of Sciences elects female member

1945 - Royal Society admits female members

1963 - First female in space, Valentina Terashkova

1990 - Society for Women's Health Research

1991 - NIH Women's Health Initiative

1993 - NIH Revitalisation Acts

2005 - Committee on Women in Neuroscience

Figure 2.1 Time line representing selected eras in the history of female education.

a record of diseases of females and children, was written as a guide for female physicians. At that time, female patients were treated only by female physicians and medical specialties were well established. Isis was the "healing goddess" of the Egyptians and the Temples of Isis were houses of healing. Within the temples, in special birth houses, the patients were cared for by female obstetricians and midwives. Hatshepsut (1503–1482 BC) championed female physicians and established three medical schools, although her brothers happily claimed the credit for her work. During this time, priestess-healers, trained in pharmacology, supervised the growing of medicinal herbs and the preparation of medications. In Greek literature of the eighth century BC, there were several mentions of female healers called "leeches." In *The Iliad*, Agamede, the daughter of Angea, was said to have cared for soldiers wounded in the battle of Troy. In *The Odyssey*, an Egyptian woman known as Polydamma was credited with supplying illicit drugs to Helen, daughter of Zeus. "Into the bowl in which their wine was mixed, she [Helen] slipped a drug that had the power of robbing grief and anger of their sting and banishing all painful memories" (Homer c. 800 BC, Book 4, 220–26). The female dynasty of Egyptian queens began around 400 BC. The queens themselves were physicians, and they enthusiastically promoted all manner of medical and scientific practice. The medical schools flourished. The physicians of classical Greece received their education in Egypt, particularly in the medical school in Alexandria. The end of the female dynasty saw the decline of the role of females in medical practice and the rise of male priests and healers in Egypt.

In Greece, the Athenian wife was seen as the child bearer and chief domestic servant. She had no independent status and could not own property, with the exceptions of clothing, jewelry, and slaves. It was even up to her husband to decide whether to keep a newborn child. Some were taught to read and write, but it is often said that during this era the most intellectually advanced Athenian females were prostitutes. In 445 BC, Pericles scandalized Athenian society by replacing his wife with Aspasia of Miletus. Aspasia was the proprietress of a house of young prostitutes who were renowned for their intellect as well as their beauty. By around 340 BC, Greek girls were educated in gynacaea (female-only schools), where they were taught the skills necessary for successful household management. Despite the low general status of females, however, there was a tradition of female healers. Theano, for example, was the wife of Pythagoras and after his death took over the running of his school, teaching philosophy and medicine. Pythias, the wife of Aristotle, coauthored a number of her husband's works for which he claimed credit, calling her his "assistant." It is clear that some Greek females were allowed to study and practice medicine, but it seems to have been with the mentorship (and probably protection) of a prominent male, usually a husband or father. One area where females did outperform males was in the Roman slave markets, where a healthy female often cost 50 times the price of a healthy male. Several important texts were written during this era. A gynecology and obstetrics text written by Aspasia in the first century AD was the standard text until the work of Trotula Platearius, in the twelfth century. Cleopatra (not the Egyptian queen, 129–201 AD) wrote a gynecology text that was in use until the sixteenth century (Brooke 1993).

Around 200 BC Greek physicians, both male and female, began practicing in Rome, and females had relatively free access to the professions until around 100 BC. During the first century BC, the status of the females in Rome was somewhat better than the status of Greek females. Roman females could own property and a dowry was not mandatory. The Roman males, however, were generally much better educated than their wives. Around 7 BC, a carpenter's wife named Mary gave birth to a son in Bethlehem and a whole new era began.

For a comprehensive discussion of the role of female healers in ancient Egypt and classical Greece, *Women Healers Through History* (Brooke 1993) is highly recommended. The rise of Christianity has often been blamed for the denial of education to females. This accusation, while partially true, certainly does not provide a complete explanation. As Christianity became the dominant religion, Christian females became physicians. One such woman, Fabiola (who died in 399 AD), was a physician who dedicated her life and practice to the care of the poor and opened the first hospital for the poor in Rome. By 385 AD, four monasteries had been established at Bethlehem. Three were for females, and the head of one of them, Eustochium, was well educated, had learned Hebrew, and edited Jerome's translation of the Bible.

There were problems for female scholars, however. In Alexandria, a female academic at the University of Alexandria was killed by a Christian mob in 415 AD, her scholarly position being viewed as against Christian dogma.

Starting around 300 AD, European society slid into the era of the "Dark Ages." Schools and libraries were destroyed and the study of medicine, mathematics, and philosophy ceased. Religious orders took over the practice and preservation of healing skills. The exception was in Celtic Britain where the study and practice of medicine continued and female physicians became part of the folklore of the British Isles. Morgan le Fay is remembered in Arthurian legend as the high priestess and healer of Avalon.

The situation in the Middle East was somewhat different. In 610 AD, the newly founded Muslim faith decreed that both females and males should receive secular and religious education. The medical school at Baghdad had 6000 male and female students.

The end of the Dark Ages around 1000 AD saw the return of scholarship to Europe. Salerno became famous as a healing center and the medical school was deemed to be the best in Europe. The school was open to males, females, Jews, and Muslims. The 1100s saw the publication of two medical books by female authors. *Diseases of Women*, by Trotula Platearius, a professor of medicine at the University of Salerno, became the major gynecology text in use for the next seven centuries. *The Book of Simple Medicine*, by Hildegard von Bingen, a German abbess, identified 47 diseases and 300 medicinal herbs. In addition to *The Book of Simple Medicine*, Hildegard wrote medical and scientific papers and music. She was also an accomplished painter. It seems clear that up until the twelfth century, despite occasional lapses, the Christian movement generally supported education and scholarship for females.

Two events occurred, however, that severely curtailed the educational prospects of females. In 1229, the Inquisition of Toulouse forbade the reading of the Bible by laypersons. This ban effectively placed European education in the monasteries and thus made it available only to males. In England, the tradition of female academia within convents remained until 1534 when Henry VIII closed all convents and schools. The closure lasted for 50 years, and, when the ban was lifted, England had been effectively deprived of its female teachers and scholars.

The first universities appeared around 1200 in Bologna, Paris, and Oxford. The universities were established as training grounds for the clergy. The teachings from classical Greece were rediscovered and formed the basis for most of the teaching and scholarship within the universities. As females could not enter the clergy, they were also banned from the universities. Although at that time females could still obtain an education within the convent system, the university ban effectively prevented females from having access to philosophy and mathematics, the foundations of modern science. In addition, because the clergy were not permitted to marry, potential female scholars were denied access to educated male mentors (Wertheim 1995). The ban on females in universities lasted for almost 700 years.

The rise of modern science

Sir Francis Bacon is often cited, at least by British and American authors, as being the father of modern scientific method. In the same breath, he is also often cited as the originator of the view of science as a masculine pursuit (Keller 1985). In his writings, Bacon used a number of metaphors for science (the male) and nature (the female). Although there is some disagreement in the literature concerning the extent of Bacon's belief in "masculine science," his true thoughts are in some ways immaterial. Bacon has served as an icon of scientific method for 400 years, widely respected and quoted. To attribute a view of "masculine" science to him has probably been, at least on some occasions, very convenient. It has also probably been used in a particular, flawed form of logical argument known as "argument from authority." Argument from authority can be a useful, and often successful, means of supporting a dubious position ("It has to be right: X said so").

What is clear is that in Bacon's time empirical investigation of scientific questions was coming into its own. By the mid-1600s, the microscope was accepted as a useful scientific tool. Physiology and biology blossomed. By the end of the century, red blood cells, capillaries, and various microorganisms had been discovered. In 1662, the Royal Society was established.

The Society, with its publication *Philosophical Transactions*, offered a forum for the discussion and publication of experimental results.

In the latter half of the 1700s, the foundations for the modern discipline of neuroscience were being laid. When in 1766 Albrecht von Haller discovered the structure and function of nerves and their connections to the brain, he uncovered the physiological foundation for the connection of the "mind" to the body. The establishment of this connection had far-reaching consequences, although ironically, even today there are philosophers and psychologists who will argue for the "mind" as a separate entity. One of the most important advances of this time was the establishment of the first insane asylums in France and the recognition of insanity as an illness rather than a derangement of the soul or possession by demons. Also, around this time, anesthetics were discovered, giving a much more humane perspective to the practice of medicine.

In 1810, Franz Gall published the first of four volumes on the structure and function of the nervous system. Much of his work was later shown to be correct. He identified the function of the gray and white matter and made a systematic analysis of the different parts of the brain. His work was sophisticated and insightful and should stand as a milestone in the study of neuroanatomy. Unfortunately, Gall went off on a tangent and tried to develop a discipline of neural analysis based upon the shape and location of bumps on the skull, "phrenology." It is sad that it is for this diversion that Gall is generally remembered.

By the mid-1800s, the formula for female neglect was well established. Few females had the education or support and encouragement to pursue any kind of science, let alone a scientific investigation of the qualities of the female brain. Females who did receive medical training often worked in areas catering to the care of females and children or the poor. In many cases, this was probably by choice, but in at least some cases these were clearly the only areas where they were allowed to practice medicine.

As the study of the brain became a legitimate scientific pursuit, more and more physicians, anatomists, and physiologists began the systematic study of its structure and function. At the same time, the pursuit of scientific knowledge left the privately funded laboratories adjoining the homes of researchers, where some females, at least, had access to the facilities. The development and maintenance of research facilities became the domain of universities and hospitals. Females, without formal scientific education, were relegated to the roles, if any, of cleaners and technicians. The few females who were qualified to conduct research would have faced an additional hurdle. When research became institutionalized, the acquisition of funding for research and the provision of facilities became competitive. Hospital or university administrators reviewed the requests for research support. They granted funds and facilities to those individuals and projects that they considered worthy of such investment. It is doubtful that a board of university administrators in the 1800s would have voted a research proposal on the structure and function of the female brain worthy of pursuit. Even today it can be difficult to convince granting bodies of the worthiness of such questions. If membership in learned societies is used to gauge female acceptance into the scientific community, then it is worth noting that the first female member of the National Academy of Sciences in the United States was elected to membership in 1925. It took the Royal Society another 20 years to allow the election of female fellows.

It is probably the case that in the earliest stages of brain research, even if females had been academically involved, the questions of sex differences would not have arisen. The essential questions were related to understanding very basic aspects of structure and function. It must not have been long, however, before behavioral differences between male and female laboratory animals became apparent, and confounding, factors in brain research. The behavior of male animals did not vary because of a breeding cycle. Female animals were in a constant, albeit predictable, state of change. Female scientists would probably have realized the significance of these behavioral differences and, just maybe, a parallel study of male and female neuroscience would have developed. Unfortunately, few female scientists were in the laboratories, at least in positions of authority, to pursue these crucial observations. The fluctuating nature of female physiology became just another laboratory problem to be overcome. The solution to the problem was easy: males became almost exclusively the experimental subjects, while females were relegated to the role of "breeders." After all, it made good financial sense. A small number of males and their harems could keep the laboratories supplied with male animals for experimentation. It is ironic that the females of experimental species followed the plight of human females, valued for their capacity to reproduce but generally barred from other activities, even participation in an experiment.

Another point for consideration was the "suitability" of females as research subjects. Females, as the weaker, "fairer" sex, were not considered robust enough to be the subjects of scientific research. Their mythical physical frailty suggested they would be more at risk from dangerous drug side-effects, or more easily exhausted by the rigors of psychophysical experiments. The well-recognized weakness associated with the menstrual cycle introduced an inconvenient variable. The problem of possible pregnancy was an even greater hurdle. At a time when pregnancy tests did not exist, any female of reproductive age was suspect. There were, in fact, two issues relating to pregnancy: willingness to administer experimental drugs to a pregnant female and ruling out pregnancy before including females. If it had been the case that the drugs being tested (or other drugs for that matter) were not administered to pregnant females, exclusion might have been justified. Once testing was complete, however, the drugs were generally administered to females, pregnant or otherwise.

Even as late as the 1960s, for example, in England drugs such as barbiturates were administered to treat hypertension in pregnancy.

To be fair, many of the laws governing medical experimentation were formulated during the time when the Nazi atrocities of WWII were a recent memory. In addition, several cases of unethical research were uncovered in the United States. The resulting public outcry led to the formulation of laws to protect the subjects of medical research. The problem was that, in an effort to protect, the structure of the laws actually led to the exclusion and disadvantage of the "minorities" that they were originally designed to protect.

With the development of the pharmaceutical industry, the policy restricting research to male animals moved from neglectful to ridiculous. Drug development and research in male laboratory animals was followed by testing using human males. This routine was so well established that the estrogen-based female contraceptive pill was tested on males! The rationale used for screening drugs in human males was delightfully illogical; females were not suitable test subjects because of their changing hormonal levels during the menstrual cycle and there was always the risk of producing birth defects. True, however, these very drugs were usually destined for use by both females and males. Ironically, when a drug-induced disaster (thalidomide in the 1960s) did occur, it was pregnant females and their children who were the victims.

Both Plato and Aristotle observed the brain of the human male to be larger than the brain of the female. Aristotle also held that the structure of the male brain was more complex, in order to allow for better "ventilation." In the 1800s, the eminent neuroanatomists Paul Broca and James Crichton-Browne debated whether the difference in brain size could be accounted for by the general difference in body size between males and females. By the final quarter of the 20th century, substantial experimental evidence had accumulated to suggest differences in brain function between females and males. From experimental psychology came research demonstrating differences in perceptual skills and motor performance (Chapter 6). From psychiatry came evidence of female to male differences in responses to antipsychotic drugs (Chapter 8). It seemed that the scientific community should have been poised to start investigating the differences.

Interestingly, it was the long overdue feminist movement that may have actually impeded progress. As feminism gathered strength and support, the emphasis was on closing the gaps between females and males in education, employment, and civil liberties. At last, opportunities for females in all aspects of life were becoming available. But the new opportunities were hard-won and many prejudices were lurking, just waiting for an invitation to resurface. It was not a politically opportune time to promote research on the very differences that so many dedicated individuals (both female and male) were striving to overcome.

Since the early 1980s, one particular researcher has pursued the question of sex differences and behavior. Over the past 25 years, Doreen Kimura has published on a number of areas where she has observed female to male differences including language (Kimura and Harshman 1984), performance on spatial tasks (Hampson and Kimura 1988), and the effects of hormone replacement therapy in protecting cognitive function after menopause (Kimura 1995). For the interested reader, "Sex Differences in the Brain," published in *Scientific American* (Kimura 1992), and *Sex and Cognition* (Kimura 1999) are highly recommended.

In the 1990s, two initiatives by the U.S. National Institutes of Health (NIH) signaled positive change for female health and health research. In 1991 the U.S. government approved the NIH Women's Health Initiative, a 14-year, $625 million research project. The aim of the initiative was to conduct multicenter studies of issues relating to the health of postmenopausal females over the period 1993 to 2007 (Thaul and Horta 1993).

In 1993, the NIH Revitalization Acts were passed (Mastroianni et al. 1994a,b). There were several stimuli for the implementation of the Acts. Pressure from "minority" groups, including females, homosexuals, and people with HIV, to be included in research initiatives had become intense. In addition, a government audit of NIH revealed that a 1986 policy of greater inclusion of females in research had not been successfully implemented and females were still "underrepresented."

The Revitalization Acts include wide-ranging recommendations for the inclusion of females in all aspects of medical research. The following are only some of the recommendations:

1. NIH is to establish a register of its research activities that includes information on participants, including sex and ethnic representation.
2. Medical and health research must benefit all people regardless of sex, race, or age.
3. Volunteers for medical research must be allowed to participate without prejudice because of sex, race, or age. In addition, the researchers must ensure that anyone who may benefit from the research is enrolled in suitable numbers to allow meaningful results. Financial concerns are not a valid reason for failure to do so.
4. Researchers must design experiments to avoid gender bias and the NIH must encourage females of all ethnic groups to pursue careers in scientific research.
5. Researchers must not exclude males or females of reproductive age, or pregnant or lactating females, from studies of drugs or procedures that will ultimately be used by these groups.

In 1990 the Society for Women's Health Research was established. The society, via its Web site (http://www.womenshealthresearch.org), offers up-to-date information and educational features for both consumers and health care professionals. One of the aims of the society is to highlight the need for the inclusion of females in research studies and the need for research on conditions that occur only or disproportionately in females, or where the expression of the condition differs between females and males. The society also sponsors meetings and other events to promote the fair inclusion of females in medical research. In 2005 the Committee for Women in Neuroscience was formed as an affiliate of the Society for Neuroscience. The International Brain Research Organization has founded the "Women in World Neuroscience" Working Group. The situation of women in academia is improving, but there is still a long way to go. "Women hold less than 15% of full professorships in Europe, even though more than half of the European student population is female" (Ledin et al. 2007).

Conclusions

Females were, most likely, the original healers. The power associated with healing, however, ultimately led to the males, in most societies, claiming the roles of healers. It is probably this change in power base that led to a lack of research into female health interests, generally, and the female brain, in particular. The points for consideration from this chapter are as follows:

1. Historically, females have been excluded from medical research on various grounds of "unsuitability" including frailty, the menstrual cycle–associated fluctuations in physiology and behavior, and the possibility of pregnancy.
2. Despite the exclusion of females from research, results obtained in males have been assumed to apply equally to females.
3. Recent NIH research initiatives have provided a legal basis for including females in all aspects of medical research.

Bibliography and recommended readings

Brooke, E. 1993. *Women Healers Through History*. London: Women's Press.
Hampson, E., and D. Kimura. 1988. Reciprocal effects of hormonal fluctuations on human motor and perceptual-spatial skills. *Behavioral Neuroscience* 102: 456–59.
Homer. C. 800 BC. *The Odyssey*, Book 4, 220–26. In E. V. Rieu, trans., rev. by D.C.H. Rieu. 1991. London: Penguin, 51.
Keller, E. F. 1985. *Reflections on Gender and Science*. New Haven: Yale University Press.
Kimura, D. 1999. Estrogen replacement therapy may protect against intellectual decline in postmenopausal women. *Hormones & Behavior* 29: 312–21.

————. 1999. *Sex and Cognition*. Cambridge, MA: MIT Press.

Kimura, D., and R. A. Harshman. 1984. Sex differences in brain organization for verbal and non-verbal functions. *Progress in Brain Research* 61: 423–41.

Ledin, A., L. Borman, F. Gannon, and G. Wallong. 2007. *EMBO Reports* 8: 983–987.

Mastroianni, A. D., R. Faden, and D. Federman, eds. 1994a. *Women and Health Research: Ethical and Legal issues of Including Women in Clinical Studies*, vol. 1. Washington, D.C.: National Academy Press.

————, eds. 1994b. *Women and Health Research: Ethical and Legal Issues of Including Women in Clinical Studies*, Vol. 2. Washington, D.C.: National Academy Press.

Thaul, S., and D. Horta, eds. 1993. *An Assessment of the NIH Women's Health Initiative*. Washington, D.C.: National Academy Press.

Trager, J. 1994. *The Women's Chronology*. New York: Holt.

Wertheim, M. 1995. *Pythagoras' Trousers: God, Physics and the Gender Wars*. New York: Random House.

chapter 3

Brain structure
The architecture of difference

The exploration of a physical system can proceed in several ways. For the structuralist, the exploration will include a detailed examination and the recording of all of the physical features available to observation. For the neuroanatomist, for example, the examination may begin with the gross anatomy, holding the preserved brain in gloved hands and inspecting the surface, then move to the microscopic anatomy, the study of cell structure using light or electron microscopy. Finally, at the molecular level, techniques such as gel chromatography can analyze the protein composition of cells, or even parts of cells.

For the functionalist, the challenge is to understand the processes performed by the structure. The neurophysiologist also uses macro- and micro-techniques in the exploration. The pupillary reflex can be observed with the naked eye. The activity of single neurons in the reflex pathway can be recorded using glass electrodes with microscopic tips. Finally, the neurotransmitters released by activity in the reflex pathway can be identified using microdialysis techniques.

Presented in this way, the distinction between structure and function seems clear. Consider what happens with the development of drug tolerance, however. Following prolonged or repeated exposure to a particular drug, the system becomes less sensitive to the drug. A larger dose is required to achieve the effect previously obtained by the smaller, original dose. This process may occur in several ways: the affinity of the receptor for the drug may decrease so that the drug is less likely to bind to the receptor; or the drug may bind with the same affinity but produce a smaller effect; or the number of receptors may decrease so that the same amount of drug has fewer binding sites and, therefore, produces less response (Figure 3.1). In all of these cases, a structural change underlies the functional change and the structure/function distinction breaks down.

The focus of this chapter will be primarily on the structure of the brain, and how it differs between the sexes. The next chapter will concentrate on the functional differences between the brains of males and females, in terms of neurochemistry. Both chapters will, of necessity, stray into the areas where the distinction breaks down and the placement of a topic in one chapter or the other becomes arbitrary.

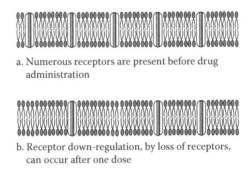

a. Numerous receptors are present before drug
   administration

b. Receptor down-regulation, by loss of receptors,
   can occur after one dose

c. Down-regulation continues with repeated doses until
   few, if any, receptors are left to respond to the drug.

Figure 3.1 Receptor down-regulation. A functional change occurs as a result of a physical change.

Development of the brain

The potential for structural differences between the female and the male brain develops early in gestation when testosterone in the developing male fetus causes the differentiation of the sex organs and forms the foundation of the female/male differences that will endure throughout life.

Human cells may exist in two states, as haploids, which carry a single set of chromosomes, and as diploids, which carry a double set of chromosomes. Cell proliferation, and therefore growth, usually occurs by mitosis, the division of the parent cell to produce two identical daughter cells. The two daughter cells each divide to produce two more cells, and so the cycle continues. For cells specialized for sexual reproduction, gametes, the process is different, however. Gametes are haploid cells: the oocyte, contributed by the female, is a nonmotile cell, containing a single X chromosome. The female always contributes an X chromosome; the female chromosomal pattern is XX, therefore, the parent cell will produce X containing gametes (Figure 3.2). In contrast to the oocyte, the sperm is a small, highly motile cell that may contain either an X- or a Y-patterned chromosome. The male parent cell, which divides to produce the individual sperm, contains an XY chromosome pattern. When the cell divides, the gametes will be either X or Y. Whether the fertilized oocyte becomes an XX, female, or an XY, male, depends entirely upon which sperm reaches and successfully penetrates the cell membrane first. At conception, the sex of the zygote (strictly speaking, it becomes an embryo at three weeks and a fetus at nine weeks)

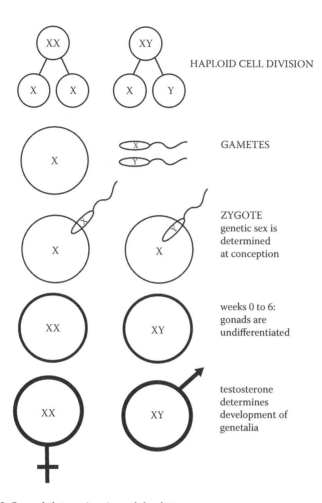

Figure 3.2 Sexual determination of the fetus.

is determined by the chromosomal pattern as either genetic female (XX) or genetic male (XY) (Alberts et al. 1994).

It is the presence or absence of the Y chromosome that determines whether the gonads, which are undifferentiated at the time of conception, will develop to be male or female. If only X chromosomes, or even only one X chromosome, are present, then the gonads will develop into ovaries. It is only if a Y chromosome is present that the gonads will develop into testes. In other words, the fetus will be female unless proven otherwise.

Until gestational week 6, the gonads are identical in both females and males. Two sets of genital ducts are present, the Müllerian ducts, which have the potential to differentiate into the female genitalia, and the Wolffian ducts, which have the potential to differentiate into male genitalia.

At postconception week 6, under chromosomal control, the gonads in the male begin to differentiate into testes, and these developing testes secrete testosterone. It is the presence of testosterone that causes the differentiation of the embryo into a male. Testosterone stimulates the Wolffian ducts to further develop into the internal sex organs, while at the same time, the peptide hormone, Müllerian inhibiting factor, also secreted by the testes, suppresses the development of the Müllerian ducts. At the same developmental time, without the addition of testosterone, the undifferentiated gonads in the female begin to differentiate into ovaries and the Müllerian ducts begin developing into female sex organs. At this stage, in both female and male, the external genitalia are also undifferentiated, and androgens secreted by the testes cause the differentiation into male genitalia. Here the role of chromosomes in sexual development finishes.

Sexual dimorphism in the CNS

The sexual differentiation of the CNS is a much slower process than the development of the sex organs. Research suggests that the development of sexual dimorphism continues well beyond birth, with spurts of development taking place at different times and in different brain regions, depending upon the sex of the individual. Indeed, it has been suggested (Swaab and Hofman 1995) that in one region specifically related to sexual behavior, the sexually dimorphic nucleus of the preoptic area, the changes continue throughout life. The variables that contribute to sexual dimorphism in the brain have been the subject of some disagreement. Until fairly recently, the general assumption has been that, as with the development of the sex organs, the presence of testosterone is the stimulus for differentiation into a male brain, and absence of testosterone results in development of a female brain. Recently, however, this idea has been challenged and evidence now suggests that the presence of estrogen and/or progesterone, not just the absence of testosterone, is necessary for the differentiation of at least some of the sexually dimorphic brain regions (Patchev et al. 1995; Wagner et al. 1998). Hormones alone, however, are not totally responsible for sexual differences, and in Chapter 4 we will explore the interactions of neurotransmitters and neuromodulators with the hormones present in the brain.

Although research on sexual dimorphism in the animal brain has arisen relatively recently, the literature on human brain dimorphism goes back about 2,000 years. As discussed in Chapter 2, both Plato and Aristotle observed that the brain of the human male was larger than the brain of the female.

Now, in the twenty-first century, there is still debate over whether there are real size differences in the brains of females and males, and what, if anything, such differences might mean. A number of authors writing on the subject state flatly that there is a 10 percent difference in the size of the female brain relative to the male brain and cite a 1978 study (Dekaban and Sadowsky 1978) to support the claim. Other, more recent, studies have reported that with adjustment for body size, female and male brains are roughly the same size. Blatter et al. (1995) have compiled tables of normative volumetric data using the MRI scans of 194 healthy adult brains of individuals ranging in age from 16 to 65 years. The authors presented the data in two forms: as volumetric estimates corrected for differences in total intracranial volume; and as uncorrected data, shown in both instances as mean volumes sorted by decade of life and sex. The authors report that age-related changes in cerebrospinal fluid volume were observed for both males and females. When the data were corrected for differences in total intracranial volume, however, most, but not all, of the sex-related differences in the volume of brain tissue disappeared. Falk et al. (1999) collected similar data from 44 female and 39 male rhesus monkeys. From their data, the authors concluded that, even when adjustments were made for body weight, male monkeys had larger brains than females of equivalent weight (Falk et al. 1999).

Whether or not there are differences in total brain volume between male and female humans, there is certainly evidence to suggest that there are differences in the volume of particular brain regions of males and females. A recent study by Reiss et al. (1996) used MRI scans from 85 children, aged between 5 and 17 years. There were 64 females (mean age 11 years) and 21 males (mean age 11 years). Handedness was matched between the two groups with 83 percent of the subjects being right-handed, 15 percent left-handed and 2 percent having no hand preference. A subset of the female (n = 57, mean age 11 years) and male (n = 12, mean age 10 years) subjects were also tested for IQ. The mean IQ for the two groups was 113 for the females and 105 for the males. All of the participants were in good health and had no known psychiatric or neurological disorders.

The scans were analyzed to determine the volume of gray matter, white matter, and CSF. After correction for intracranial volume, the results of the study revealed significant differences in brain volume attributable to sex. The total cerebral volume (gray matter + white matter + CSF) was approximately 10 percent larger in males than in females. The difference was primarily due to an increased volume of cortical gray matter in males (and, to a lesser extent, to a larger volume in the lateral ventricles). There were no differences in the volume of the caudate nucleus, lenticular nucleus (i.e., the putamen + the globus pallidus), or thalamus (Table 3.1). Although there were no significant differences directly attributable to age, the authors comment that with increasing age there was a trend for decreasing volume of gray matter and increasing volume of white matter. Further analysis of the data revealed that the volume of gray matter in the prefrontal regions was related to the IQ of the subjects. Smith and colleagues (2006) examined the relationship between age and sex in an MRI study using 122 healthy elderly people (female = 71, males = 51) aged

Table 3.1 Brain volume estimates in children using MRI.

Measure	Result
Absolute brain volume	m > f
Right hemisphere volume	m > f
Left hemisphere volume	m > f
Total gray matter volume	m > f
Cortical gray matter	m > f
Caudate nucleus gray matter	m = f
Lenticular nucleus gray matter	m = f
Thalamus gray matter	m = f
White matter	m > f
CSF	m = f
CSF outside ventricles	m = f
CSF in lateral ventricles	m > f

Source: Adapted from Reiss et al. (1996), 1766.

between 58 and 95 years with a mean age of 75 years. The authors reported that gray matter volume decreased at a rate of 2.4 cm³/year while CSF volume increased by 2.5 cm³/year. There were diffuse age-related changes in gray matter volume found in cortical regions, the cerebellum and the basal ganglia. A specific reduction in white matter was also reported in the anterior region of the corpus callosum. Interestingly, they did not observe any gender-specific effects.

In a similar study, Giedd et al. (1996) used MRI to study the brain structure of 104 children aged between 4 and 18 years. The authors reported that males had significantly larger cerebral (9 percent) and cerebellar (8 percent) volumes than females. Regional analysis showed a larger putamen and globus pallidus in males but relatively larger caudate nucleus in females. Also in agreement with the previous study, no age-related changes in the cerebral or cerebellar volumes were observed. Several regional changes were observed to occur with age, however. In males only, a significant increase in the volume of the lateral ventricles and a significant decrease in the volumes of the caudate and putamen were observed, suggesting that sex-specific maturational changes in the volume of specific brain regions were occurring.

Andreasen et al. (1993) looked for correlations between brain volume, gender, and IQ in adults. The subjects for the study were 67 normal, healthy adults recruited through newspaper advertisements. There were 30 female subjects (mean age 38 years, mean height 168 cm) and 37 males (mean age 38 years, mean height 181 cm). Regional and total brain volumes were measured using MRI; intelligence was measured using the Wechsler Adult Intelligence Scale — Revised (WAIS-R). The WAIS-R is an instrument for assessing adult intelligence that gives an overall IQ score, as well as individual scores for Performance IQ and for Verbal IQ. The mean full-scale IQ for all subjects was 116, the mean verbal IQ was 114, and the mean performance IQ was 114. There were no significant differences between female and male IQ. The MRI scans were analyzed for the volume of different brain regions, with discrimination made between gray matter, white matter, and CSF. The MRI data, corrected for body-height differences, were then compared to the IQ data for correlations between regional volumes and IQ.

The results of this study showed that, for both male and female subjects, there was an overall positive correlation between brain volume and IQ, with all of the differences being attributable to the volume of the gray matter. In addition, there were male-female differences in the correlations of the volume of specific brain regions and Verbal and Performance IQ. For females, there were correlations between Verbal IQ and total brain volume, the volumes of the cerebellum, the cerebral hemispheres, the left and right temporal lobes, and the hippocampi. Performance IQ, on the other hand, was correlated only with the volume of the cerebellum, the right temporal lobe, and the left hippocampus. Results for the male subjects showed correlations between Performance IQ and the total brain volume, and the individual volumes of the cerebral hemispheres, the cerebellum, and the left hippocampus. For Verbal IQ the only correlations were for the volumes of the right temporal lobe and the cerebellum (Table 3.2).

Table 3.2 Correlations between brain volume and IQ.

Structure	Sex	V/IQ	P/IQ	FS/IQ
Total cranium volume	m	-	*	*
	f	*	-	*
Right hemisphere volume	m	-	*	*
	f	*		*
Left hemisphere volume	m	-	*	*
	f	*	-	*
Right temporal lobe volume	m	*	-	*
	f	*	*	*
Left temporal lobe volume	m	-	-	-
	f	*	-	*
Cerebellum volume	m	*	*	*
	f	*	*	*
Right hippocampus volume	m	-	-	-
	f	*	-	*
Left hippocampus volume	m	-	*	*
	f	*	*	*

Source: From Andreasen et al. (1993). FS/IQ, full scale IQ; P/IQ, performance IQ; V/IQ, verbal IQ.
* *Indicates significant positive correlation.*

This study is interesting for several reasons. First, there is a clear indication that "the larger the brain, the higher the IQ" (p. 132), although, as the authors are quick to point out, the differences are "modest." Therefore size is only one of a number of factors contributing to IQ. More interesting, however, is the finding that the correlation between volume of different brain regions and particular types of IQ, Performance or Verbal, differs between females and males.

Regional differences

Differences between females and males have been reported for several cortical structures. A postmortem study of the language-associated cortical areas of the brains of 11 females and 10 males has reported that the volume of the superior temporal cortex was 18 percent larger in females than in males. The overall difference in volume was accounted for primarily by differences in the planum temporale (mostly the auditory association cortex), which was 30 percent larger in females (Harasty et al. 1997).

A cytoarchitectural study of the planum temporale in females and males has reported particularly interesting results (Witelson et al. 1995). This study offers some especially useful information because although it used tissue obtained postmortem, not only was the cause of death well understood but the subjects were also known to have had normal cognitive function in life. The subjects for the study were five females and four males suffering from metastatic cancer. At the time the subjects were recruited for the study they were free from obvious symptoms of their disease. The subjects gave their consent for their brains to be examined at autopsy. At the time of recruitment, they underwent extensive medical and neuropsychological testing. In addition, they developed no neurological or psychiatric complications as the disease progressed. All subjects were predominantly right-handed. The mean age at death was 49 years for the males and 54 years for the females.

Following death, the brains were removed and fixed for histological analysis. The time between death and tissue fixation was on average 3.5 hours for the males and 7.2 hours for the females. The variables measured were the total cortical depth, the number of neurons per 1 mm^2 of cortical surface and the number of neurons per unit volume (the density of cell packing). Neither the cortical depth nor the number of cells differed between females and males. The number of neurons per unit volume, however, was significantly greater in females (by 11 percent) than in males. The authors point out that this is almost the same difference as the difference in total brain volume generally reported between females and males. They suggest that reported overall differences in brain volume may simply be because cells are packed more densely in the female brain than in the male brain. A similar suggestion has been made by Gur et al.

(1999), who have reported, using MRI, that females have a higher percentage of cortical gray matter while males have a higher percentage of cortical white matter. They suggest that the increased density of gray matter in females could compensate for the smaller overall volume. By contrast, Rabinowicz et al. (1999) have reported significantly higher neuronal density and higher estimated neuronal numbers in males than in females.

Despite superficial appearances, the brain is not symmetrical around its longitudinal axis. This asymmetry has been shown to apply to both structure, discussed in this chapter, and function (Chapter 4). Many studies have demonstrated that the brains of normal individuals are asymmetrical, with the right hemisphere larger than the left, and that this asymmetry is evident in the newborn child. This normal asymmetry is so well documented, in fact, that individual variations in asymmetry may be indicative of brain pathology such as schizophrenia (Chapter 7). It is also reported that the female brain is more symmetrical than the male brain. The asymmetry is reported to differ most between females and males in areas related to language function and to handedness (Chapter 6).

A number of recent studies have demonstrated that, in people who are right-handed, the frontal areas of the brain tend to be larger on the right side. Paus et al. (1996) have reported frontal lobe differences in an MRI study of 105 right-handed individuals (Table 3.3). In their study, these authors reported that the volume of gray matter in the anterior cingulate sulcus and in the superior-rostral sulcus was larger in the right hemisphere than the left hemisphere, while the volume of gray matter in the posterior cingulate sulcus and paracingulate sulcus was greater on the left. The cingulate sulcus was significantly larger in females than in males, and the paracingulate sulcus was significantly larger in males. However, an imaging study by Bullmore et al. (1995) has demonstrated that these differences are not nearly so clear-cut when you consider handedness as a variable. In their study, the only consistent pattern of asymmetry was for right-handed males. As the authors point out, their result is consistent with the idea that the female brain is more symmetrical

Table 3.3 Estimated frontal lobe asymmetry using MRI.

Region	Asymmetry
Volume of gray matter, anterior cingulate sulcus	R > L, f and m
Volume of gray matter, posterior cingulate sulcus	L > R, f and m
Volume of gray matter, paracingulate sulcus	L > R, f and m
Sex differences in asymmetry:	
cingulate sulcus	f > m
paracingulate sulcus	m > f

Source: From Paus et al. (1996).

than the male brain. On the other hand, Reiss et al. (1996), in a study of children, reported that the brain asymmetry, greater volume in the right cortical areas, was the same for boys and girls. Although the proportion of left-handed to right-handed children was balanced between the two groups (15 percent were left-handed), the results were not analyzed for differences in handedness.

One area where clear size and shape differences have been noted is in the fiber tracts that connect the cerebral hemispheres. These fiber crossings are known as commissures or decussations, depending upon their location. The largest of the commissures is the corpus callosum (Figure 3.3), which relays information between the two cerebral hemispheres. Most fiber tracts in the human brain and spinal cord are bilaterally crossed and symmetrical (Figure 3.4). Sensory information from one side crosses to the opposite side and is relayed to the cerebral cortices for processing. The corpus callosum has generally been reported to be larger in females, and some authors have attributed sex differences in the lateralization of cognitive and language functions to this size difference (Chapter 6). Some studies, both histological (de Lacoste-Utamsing and Holloway 1982) and MRI (Allen et al. 1991), have suggested that the previously reported sex differences may be due to a different *shape* in females rather than a difference in the overall size of the corpus callosum. The anterior commissure, which crosses the deep diencephalon, near the hypothalamus, contains axons that primarily connect the two temporal lobes. In a study of autopsy

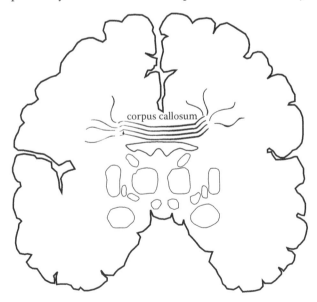

Figure 3.3 The corpus callosum connects the two cerebral hemispheres.

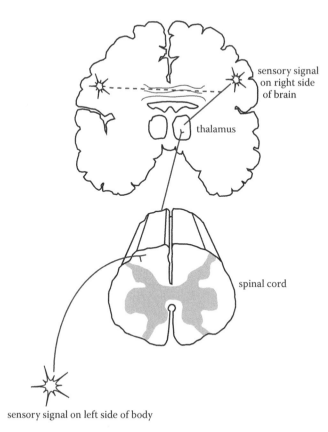

sensory signal
on right side
of brain

thalamus

spinal cord

sensory signal on left side of body

Figure 3.4 Sensory information, for example, touch, from one side of the body is relayed, via the thalamus, to the opposite side of the brain. The information is then "shared" via the corpus callosum with the other side of the brain.

material from 100 human females and males it has been reported that the anterior commissure is approximately 12 percent larger in females than in males (Allen and Gorski 1991).

The hypothalamus is known to play a role in sexual behavior and reproduction. Not surprisingly, it was one of the first regions to be systematically examined for sexual dimorphism. The hypothalamus may be divided into a number of distinct regions, based upon the function of the neurons located in each different area. Some authors classify the preoptic area as a part of the hypothalamus and others classify it as a separate nucleus. The two areas are so intimately linked, however, that to consider them together is probably the logical option, at least for our purposes (Figure 3.5).

Sexual dimorphism in the human preoptic nucleus was first reported in 1985 by Swaab and Fliers. In that study both the size and the

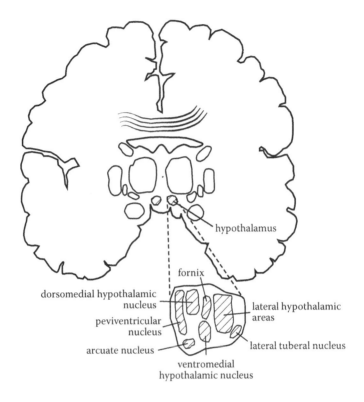

Figure 3.5 Schematic representation of the subregions of the human hypothalamus. Shaded region represents approximate location of enlarged area; i.e., internal capsule.

total number of neurons in particular regions of the preoptic nucleus were reported to be significantly larger in males than in females. The volume of the bed nucleus of the stria terminalis in males has also been reported to be approximately twice the volume in females (Allen et al. 1990; Zhou et al. 1995). The evidence for sexual dimorphism in the hypothalamus was reviewed by Swaab and Hofman (1995). In their article, the authors noted that the nucleus is the same size in females and males at birth, but is smaller in adult females relative to adult males because the size decreases with maturity in females. Another area where the female/male size relationship changes with development is the amygdala. At birth, it is the same size in females and males, but by the age of around 8 years it has grown to a larger overall size in males. At around 10 years, however, the volume has begun to decrease in males, but continues to increase in females so that from about the age of 20 onward, the volume of the amygdala is greater in females than in males (Goldstein et al. 1999).

Evidence from other species

The majority of animal studies on sexual dimorphism in the brain have come from studies in the laboratory rat, although monkeys, ferrets, gerbils, and squirrels have also had their day. In general, the results of the experimental studies are consistent with the results obtained in humans. For this reason, this section will give only a brief overview of the animal literature.

Sexual dimorphism has, not surprisingly, been clearly demonstrated in the hypothalamus. One particular region of the preoptic area has been reported to be five to six times greater in volume in male rats than in female rats (Gorski et al. 1978). This region was named the "sexually dimorphic nucleus of the preoptic area" by Gorski et al., who first reported the size difference. Another part of the preoptic area, the anteroventral periventricular nucleus of the preoptic area, shows the opposite pattern of sexual dimorphism: the area is substantially larger and more densely packed with neurons in the female than in the male.

Brain regions where anatomical dimorphism has been observed are summarized in Table 3.4. This table is not meant to be a comprehensive listing of the literature. It is meant only as a list of examples. In most cases, the difference reported seems to be a larger area in the male. One area that has been deliberately omitted from the list is that of the song-related nuclei that show distinct differences in a number of bird species. The song-related nuclei play an important role in the sexual behavior of many birds. The song of the male is part of an elaborate courtship display necessary to seduce the female. While human males may indeed use song as part of their courtship display (have a look at a rock star in action), anything even close to the song-related nuclei is yet to be discovered in the brain of the human male.

In the cerebral cortex, differences have been reported in regions related to vision, motor control, sense of smell, and emotion. In the visual cortex, there are more neurons in males than in females and the dendrites of some neurons are larger in males. Sex-based differences have been reported in the hippocampus and amygdala. In the hippocampus, which is clearly divided into layers composed of different types of neurons, the granule cell layer is larger in males than in females. The amygdala is also larger in males and receives a greater number of axons from neurons in the hypothalamus carrying the hormone vasopressin. In the brain stem, differences have been noted in the striatum and the locus coeruleus. In both of these regions, neurons containing the major neurotransmitter of the region (noradrenaline in the case of the locus coeruleus, and GABA in the case of the striatum) are more numerous in females than in males. In addition to areas containing neuronal cell bodies, a number of animal studies have also reported size differences in the corpus callosum and the anterior commissure, similar to those already discussed in humans.

Table 3.4 Examples of studies reporting regional brain differences in rat.

Region	Difference	Reference
amygdala	volume, m > f	Mizukami et al. 1983
bed nucleus olfactory tract	volume, m > f cell count, m > f	Collado et al.
corpus callosum	unmyelinated axons, f > m myelinated axons, m > f	Mack et al. 1995.
hippocampus	granule cell number, m > f	Roof 1993
	CA3 pyramidal cells, proximal dendrite volume, f > m; distal dendrite volume, m > f	Juraska et al. 1989
hypothalamus:		
anterior ventral	volume, f > m	Simerly et al. 1985
periventricular preoptic area	cell count, f > m	Leal et al. 1998
arcuate nucleus	volume, m > f, density denditric branching and spines, f > m	
medial preoptic area	volume, m > f	Gorski et al. 1978
locus coeruleus	noradrenaline neurons number, f > m volume, f > m	Guillamon et al. 1988
neocortex	volume, m > f	Reid & Juraska 1992b
bed nucleus stria terminalis	volume, m > f	De Vries et al. 1994
striatum	GABA neurons number, f > m	Ovtscharoff et al. 1992
suprachiasmatic nucleus	spine synapses, m > f	Guldner 1982
visual cortex	volume, number of neurons, m > f	Reid & Juraska 1992a

The CT scan is a multiple X-ray method that provides a snapshot of brain structure at a given point in time. The machinery is rotated around the head of the subject, taking X-rays at every degree of rotation. By summing the density of the image, for each individual exposure, for individual coordinates throughout the brain, a three-dimensional representation of the brain structures can be created that distinguishes gray matter and white matter, as well as CSF and blood. The advent of CT was a major step forward in providing anatomical analysis of the human brain. It allowed visualization of individual brain structures without the distortion associated with anatomical studies of fixed tissue. It was also a breakthrough as a diagnostic tool, providing accurate localization (about 1 mm resolution) of tumors and lesions.

An early CT study of 75 males and 41 females reported a significantly greater ratio of ventricles to brain volume in males than in females, with larger sulci and fissures in the males (Jacobson 1986). An MRI study in 1999 presented similar results. The study, which measured volume of gray matter, white matter, and CSF, found a higher percentage of gray matter in females than males. The males, on the other hand, had a higher percentage of white matter and CSF. There were no hemispheric asymmetries in females. However, gray matter percentage was higher in the left hemisphere of males, while CSF was higher in the right (Gur et al. 1999). Im et al. (2008) also concluded that reported sex differences in brain size between females and males required differences in connectivity, as suggested by the differences in gray and white matter.

MRI, like CT, also uses computerized analysis of sequential brain images to produce a three-dimensional image of the brain. "Magnetic resonance" refers to the oscillations observed in the nuclei of elements with odd atomic numbers (e.g., hydrogen, H) when they are exposed to a magnetic field. The oscillating nuclei align themselves with the magnetic field. Once the nuclei are aligned, a burst of radio waves will disrupt the alignment. When the radio waves stop, the nuclei will "snap back" into their alignment with the magnetic field, and in doing so will emit a radio signal of their own. This radio signal may be recorded, using the same multiple exposure, rotating-image protocol, to build up a three-dimensional image of the object. The human body is ideal for this kind of imaging, because the different structures have quite different chemical compositions and different water (H_2O) content. Bone, for example, contains little water so there are few H nuclei to oscillate, producing a very weak signal. White matter, on the other hand, has a higher water content and produces stronger signals. The result is a high-resolution image (about 0.1 mm). Because changes in water content accompany neuropathologies such as tumors or multiple sclerosis, MRI is a much better diagnostic tool than CT.

MRI has proved to be a valuable tool in studying the effects of aging on the living brain. The corpus callosum (CC), because of its importance to the transfer of information between the hemispheres, has received a good deal of attention. Probably the most interesting result to come out of these studies is the demonstration that while the CC has a tubular shape in males, in females it has a clearly bulbous shape at the posterior end (Allen et al. 1991). The volume of the CC has also been demonstrated to decrease with age in females. Cowell et al. (1994) have reported that the frontal and temporal lobes have a greater volume in males than in females, with the right volume being the greatest. The study of 130 subjects, 70 males and 60 females, included an age range from 18 to 80 years. The frontal and temporal lobes have also been reported to decrease with age, with greater age-related reductions in females than males (Cowell et al. 1992). A recent study of the putamen, an area related to motor function, has demonstrated that compared with females, males show a greater age-related loss in gray matter (Nunneman et al., in press).

Structural changes in the adult brain

Until 1962 the idea that the volume and structure of the adult brain could change in any way other than reductions in volume due to damage, age, or disease was regarded as a naïve impossibility (see Galea 2008 for a review). Neurogenesis (the birth and survival of new neurons) was thought to be a developmental phenomenon that ceased when development ended. In 1962, Altman (in Galea 2008) reported finding neurogenesis in the adult dentate gyrus. Since that first report, it has been shown that in rats and mice, neurogenesis in the hippocampus is related to increased learning and memory. The hippocampus is rich in receptors for estrogen, testosterone, and glucocorticoids, suggesting a possibility for sex differences in the process. In 2008 it was reported that administration of estradiol to female and male rats for 15 days resulted in sex-dependent effects on neurogenesis in the dentate gyrus of the hippocampus (Baker and Galea 2008). As expected from previous studies, hippocampal volume was greater in the male rats than in the female rats. In females, there was a higher density of proliferating cells but decreased cell survival and a lower rate of cell death. In male rats, estradiol administration had no significant effects. Eating disorders are more prevalent in females that males. It has been reported that anorexia nervosa is associated with a decreased gray matter in the anterior cingulated cortex and that the gray matter loss is directly correlated to the severity of the disorder (Muhlau et al. 2007).

Conclusions

When you consider the vast amount of data available on neuroanatomy, both in humans and experimental animals, the literature on female-male differences seems meager indeed. There are a number of possible explanations for the scarcity of hard evidence; perhaps the most compelling explanation is in terms of the technology required to find consistent structural differences in brain tissue.

It must be assumed that the differences in question will be fairly small and discreet by the standards of gross anatomy. If they were not, head sizes could be expected to differ radically between the sexes (all that extra volume would have to go somewhere) or at least the head shape might differ noticeably. It is really only in the last 20 years that scanning techniques for examining living tissue have become readily available, and even more recently that such techniques have been accessible for research purposes. Before the advent of scanning techniques, neuroanatomists were dependent upon techniques using preserved brain tissue that, particularly in the case of human brain, may have undergone structural changes, especially shrinkage, before and during the preservation procedure. In such conditions, subtle differences could be difficult to find.

Another possibility is that few researchers actually looked for sex differences. As discussed in the two previous chapters, sex differences in neuroanatomical studies may have been "averaged out" by pooling data from a mixed sex population of experimental animals, or excluded initially by using only one sex. Systematic studies comparing tissue from a number of brains have, until the advent of histological video analysis in the 1990s, been time-consuming, eye-straining work. In older studies where sexual dimorphism has been reported in areas where it would not be expected, you can only assume that painstaking work by dedicated anatomists revealed unexpected results. It is also possible that in some cases, sex differences were observed but not reported simply because they were beyond the scope of the particular study.

Despite the relative shortage of data in this chapter, it has been established that some reliable differences in architecture do exist between the brains of males and females. This architecture forms the foundation for the next chapter, in which the evidence for functional differences in terms of neurotransmitters, neuromodulators, and their interactions with hormones will be examined.

The following summarizes the main points to be drawn from Chapter 3:

1. There are structural differences between the brains of females and males in areas not directly associated with sexual function or behavior.
2. Structural differences in cortical areas may provide the foundation for observed differences in some cognitive abilities.
3. The brains of both females and males are normally asymmetrical and there are sex differences in the extent of this normal asymmetry.

A final note: When most authors discuss the functional significance of sex differences in brain structure they conclude that the structural differences probably underlie the observed functional differences. DeVries and Boyle (1998) have offered an alternative view. Perhaps, they suggest, sexual dimorphism in brain structure allows females and males to display similar behaviors despite their physiological and neurochemical differences.

Bibliography and recommended readings

Alberts, B., D. Bray, J. Lewis, M. Raff, K. Roberts, and J. D. Watson. 1994. *Molecular Biology of the Cell*. 3rd ed. New York: Garland.

Allen, L. S., and R. A. Gorski. 1990. Sex difference in the bed nucleus of the stria terminalis of the human brain. *Journal of Comparative Neurology* 302: 697–706.

———. 1991. Sexual dimorphism of the anterior commissure and massa intermedia of the human brain. *Journal of Comparative Neurology* 312: 97–104.

Allen, L. S., M. F. Richey, Y. M. Chai, and R. A. Gorski. 1991. Sex differences in the corpus callosum of the living human being. *Journal of Neuroscience* 11: 933–42.

Andreasen, N. C., M. Flaum, V. Swayze II, D. S. O'Leary, R. Alliger, G. Cohen, J. Ehrhardt, and W.T.C. Yuh. 1993. Intelligence and brain structure in normal individuals. *American Journal of Psychiatry* 150, 130–34.

Aristotle. 384–322 BC. In J. L. Ackrill, trans. and ed. 1987. *A New Aristotle Reader*. Princeton: Princeton University Press, 248–49.

Baker, J. M., and L.A.M. Galea. 2008. Repeated estradiol administration alters different aspects of neurogenesis and cell death in the hippocampus of female, but not male, rats. *Neuroscience* 152: 888–902.

Blatter, D. D., E. D. Bigler, S. D. Gale, S. C. Johnson, C. V. Anderson, B. M. Burnett, N. Parker, S. Kurth, and S. D. Horn. 1995. Quantitative volumetric analysis of brain MR: normative database spanning 5 decades of life. *American Journal of Neuroradiology* 16: 241–51.

Bullmore, E., M. Brammer, I. Harvey, R. Murray, and M. Ron. 1995. Cerebral hemispheric asymmetry revisited; effects of handedness, gender and schizophrenia measured by radius of gyration in magnetic resonance images. *Psychological Medicine* 25: 349–63.

Collado, P., A. Guillamón, A. Valencia, and S. Segovia. 1990. Sexual dimorphism in the bed nucleus of the accessory olfactory tract in the rat. *Developmental Brain Research* 56: 263–68.

Cowell, P. E., L. S. Allen, N. S. Zalatimo, and V. H. Denenberg. 1992. A developmental study of sex and age interactions in the human corpus callosum. *Developmental Brain Research* 66: 187–92.

Dekaban, A. S., and D. Sadowsky. 1978. Changes in brain weights during the span of human life: relation of brain weights to body heights and body weights. *Annals of Neurology* 4: 345–56.

De Lacoste-Utamsing, C., and R. L. Holloway. 1982. Sexual dimorphism in the human corpus callosum. *Science* (Washington) 216: 1431–32.

De Vries, G. J., and P. A. Boyle. 1998. Double duty for sex differences in the brain. *Behavioural Brain Research* 92: 205–13.

De Vries, G. J., Z. Wang, N. A. Bullock, and S. Numan. 1994. Sex differences in the effects of testosterone and its metabolites on vasopressin messenger RNA levels in the bed nucleus of the stria terminalis of rats. *Journal of Neuroscience* 14: 1789–94.

Falk, D., N. Froese, D. S. Sade, and B. C. Dudek. 1999. Sex differences in brain/body relationships of rhesus monkeys and humans. *Journal of Human Evolution* 36: 233–38.

Galea, L.A.M. 2008. Gonadal hormone modulation of neurogenesis in the dentate gyrus of adult male and female rodents. *Brain Research Reviews* 57: 332–41.

Giedd, F. N., J. W. Snell, N. Lange, J. C. Rajapakse, B. J. Casey, P. L. Kozuch, A. C. Vaituzis, Y. C. Vauss, S. D. Hamburger, D. Kaysen, and J. L. Rapoport. 1996. Quantitative magnetic resonance imaging of human brain development: ages 4 – 18. *Cerebral Cortex* 6: 551–60.

Goldstein, J. M., D. N. Kennedy, and V. S. Caviness. 1999. Sexual dimorphism. *American Journal of Psychiatry* 156: 352.

Gorski, R. A., J. H. Gordon, J. E. Shryne, and A. M. Southam. 1978. Evidence for a morphological sex difference within the medial preoptic area of the rat. *Brain Research* 148: 333–46.

Guillamon, A., M. R. de Blas, and S. Segovia. 1988. Effects of sex steroids on the development of the locus coeruleus in the rat. *Brain Research* 468: 306–10.

Guldner, F. H. 1982. Sexual dimorphisms of axo-spine synapses and postsynaptic density material in the suprachiasmatic nucleus of the rat. *Neuroscience Letters* 28: 145–50.

Gur, R. C., B. I. Turetsky, M. Matsui, M. Yan, W. Bilker, P. Hughett, and R. E. Gur. 1999. Sex differences in brain gray and white matter in healthy young adults: correlations with cognitive performance. *Journal of Neuroscience* 19: 4065–72.

Harasty, J., K. L. Double, G. M. Halliday, J. J. Dril, and D. A. McRitchie. 1997. Language-associated cortical regions are proportionally larger in the female brain. *Archives of Neurology* 54: 171–76.

Im, K., J-M. Lee, O. Yttelton, S. H. Kim, A. C. Evans, and S. I. Kim. 2008. Brain size and cortical structure in the adult brain. *Cerebral Cortex* 18: 2181–91.

Jacobson, R. 1986. The contributions of sex and drink history to the CT brain scan changes in alcoholics. *Psychological Medicine* 16: 547–59.

Jaraska, J. M., J. M. Fitch, and D. L. Washburne. 1989. The dendritic morphology of pyramidal neurons in the rat hippocampal CA3 area. II. Effects of gender and the environment. *Brain Research* 479: 115–19.

Leal, S., J. P. Andrade, M. M. Paula-Barbosa, and M. D. Madeira. 1998. Arcuate nucleus of the hypothalamus: effects of age and sex. *Journal of Comparative Neurology* 401: 65–88.

Mack, C. M., G. W. Boehm, A. S. Berrebi, and V. H. Denenberg. 1995. Sex differences in the distribution of axon types within the genu of the rat corpus callosum. *Brain Research* 697: 152–56.

Mizukami, S., M. Nishizuka, and Y. Arai. 1983. Sexual difference in nuclear volume and its ontogeny in the rat amygdala. *Experimental Neurology* 79: 569–75.

Muhlau, M., C. Gaser, R. Ilg, B. Conrad, C. Leibl, M. H. Cebulla, H. Backmund, and M. Gerlinghoff. 2007. Gray matter decrease of the anterior cingulated cortex in anorexia nervosa. *American Journal of Psychiatry* 164: 1850–57.

Nunnemann, S., A. M. Wohlschläger, R. Ilg, C. Gaser, T. Etgen, B. Conrad, C. Zimmer, and M. Mühlau. (in press). Accelerated aging of the putamen in men but not in women. *Neurobiology of Aging*.

Ovtscharoff, W., B. Eusterschulte, R. Zienecker, I. Reisert, and C. Pilgrim. 1992. Sex differences in densities of dopaminergic fibers and GABAergic neurons in the prenatal rat striatum. *Journal of Comparative Neurology* 323: 299–304.

Patchev, V. K., S. Hayashi, C. Orikasa, and O. F. Almeida. 1995. Implications of estrogen-dependent brain organization for gender differences in hypothalamo-pituitary-adrenal regulation. *FASEB Journal* 9: 419–23.

Paus, T., N. Otaky, Z. Caramanos, D. MacDonald, A. Zijdenbos, D. D'Avirro, D. Gutmans, C. Holmes, F. Tomaiuolo, and A. C. Evans. 1996. In vivo morphometry of the intrasulcal gray matter in the human cingulate, paracingulate and superior-rostral sulci; hemispheric asymmetries, gender differences and probability maps. *Journal of Comparative Neurology* 376: 664–73.

Rabinowicz, T., D. E. Dean, J.M.-C. Petetot, and G. M. de Courten-Myers. 1999. Gender differences in the human cerebral cortex: more neurons in males, more processes in females. *Journal of Child Neurology* 14: 98–107.

Reid, S. N., and J. M. Juraska. 1992a. Sex differences in neuron number in the binocular area of the rat visual cortex. *Journal of Comparative Neurology* 321: 448–55.

———. 1992b. Sex differences in the gross size of the rat neocortex. *Journal of Comparative Neurology* 321: 442–47.

Reiss, A. L., M. T. Abrams, H. S. Singer, J. L. Ross, and M. B. Denckla. 1996. Brain development, gender and IQ in children. A volumetric imaging study. *Brain* 119: 1763–74.

Roof, R. L. 1993. The dentate gyrus is sexually dimorphic in prepubescent rats: testosterone plays a significant role. *Brain Research* 610: 148–51.

Simerly, R. B., L. W. Swanson, and R. A. Gorski. 1985. The distribution of monoaminergic cells and fibers in a periventricular preoptic nucleus involved in the control of gonadotropin release: immunohistochemical evidence for a dopaminergic sexual dimorphism. *Brain Research* 330: 55–64.

Smith, C. D., H. Chebrolu, D. R. Wekstein, F. A. Schmitt, W. R. Markesber. 2006. Age and gender effects on human brain anatomy: a voxel-based morphometric study in healthy elderly. *Neurobiology of Aging* 8: 1075–87.

Swaab, D. F., and E. Fliers. 1985. A sexually dimorphic nucleus in the human brain. *Science* (Washington) 228: 1112–14.

Swaab, D. F., and M. A. Hofman. 1995. Sexual differentiation of the human hypothalamus in relation to gender and sexual orientation. *Trends in Neuroscience* 18: 264–70.

Wagner, C. K., A. Y. Nakayama, and G. J. DeVries. 1998. Potential role of maternal progesterone in the sexual differentiation of the brain. *Endocrinology* 139: 3658–61.

Witelson, S. F., I. I. Glezer, and D. L. Kigar. 1995. Women have greater density of neurons in posterior temporal cortex. *Journal of Neuroscience* 15: 3418–28.

Zhou, J.-N., M. A. Hofman, L.J.G. Gooren, and D. F. Swaab. 1995. A sex difference in the human brain and its relation to transsexuality. *Nature* (London) 378: 68–70.

chapter 4

Functional differences
Neurotransmitters and neuromodulators

This chapter begins on the other side of the arbitrary dividing line between structure and function described in Chapter 3. The distinction between structure and function is blurred when the topic for discussion is receptors, the binding sites for neurotransmitters and hormones. The receptors themselves are structures, complexes of protein units embedded in the plasma membrane of the cell, or in some cases, in the cytoplasma of the cell. Receptors are by nature plastic, however. They may change in number, distribution, or sensitivity depending upon the physiological state of the organism as well as previous exposure to drugs, hormones, and neurotransmitters. Because of their constantly changing nature, even descriptions of the structural components of receptors are more suited to a chapter on function.

Receptors are the functional units of action for neurotransmitters, hormones, and drugs. Brain cells are well-protected, self-contained units. The plasma membrane separates the intracellular space from the extracellular space, isolating the cellular machinery from environmental (extracellular) events. For a drug to influence the activity of a cell, it must have access to the intracellular signaling mechanisms, and, for the majority of drugs, the receptor is the point of access. It is logical that the investigations of drug actions in the brain begin with the search for, and identification of, drug-specific receptors. The study of hormone actions in the brain is no exception.

In 1968, McEwen et al. discovered that when minute doses of radioactively labeled steroids were injected into the bloodstream of rats, the steroids were selectively taken up by certain regions of the brain. The resulting areas of radioactivity indicated the presence of specific binding sites, receptors, for the steroids. This experiment changed forever our perspective on sex differences in the brain. The discovery of steroid receptors, in this case for corticosteroids and progesterone (McEwen et al. 1968), meant that the hormones normally found in the circulating blood of females and males could cross the blood brain barrier and act directly on brain cells. Changes in circulating hormone levels could, potentially, be reflected in changes in neuronal activity and, therefore, behavior.

Long before the search for the steroid/hormone receptors began, certain areas in the brain associated with sexual function had been identified. Some of these early studies were crude. Electrical stimulation, using metal electrodes implanted in the area of interest, was used to study the effects of "activating" a specific area on the behavior of the animal. Alternately, specific areas were lesioned, using either electric current or surgical techniques, and the effect on behavior observed.

As a result of this early work, many of the regions of the brain associated with sexual function had been identified. These areas, particularly the hypothalamus and limbic system, served as the starting place in the search for hormone receptors.

In the ten years following McEwen's initial experiments, many other studies confirmed and extended the original results (see McEwen et al. 1986 for a review). Binding sites for corticosteroids, androgens, estrogens, and subtypes of each were identified. As the different types of receptors were characterized, it became clear that they had very different patterns of distribution within the brain. By the early 1990s, comprehensive maps of steroid receptor binding sites in the rat brain had been published.

Not surprisingly, the greatest density of estrogen binding sites was identified in the amygdala and areas associated with control of reproduction, the preoptic area, the bed nucleus of the stria terminalis and the hypothalamus. The distribution of binding sites for progesterone was very similar to the distribution of estrogen binding sites. The greatest density of testosterone receptors has been shown in the hypothalamus, preoptic area, the bed nucleus of the stria terminalis, amygdala, septum, and the hippocampus. In addition, binding sites for both hormone types have been demonstrated, albeit in very low density, in other brain regions including the brain stem and cerebellum.

Estrogen, progesterone, androgens, and their receptors in the CNS

An interesting characteristic of steroids is their ability to cross cell membranes. In the majority of cases, drugs and neurotransmitters will not cross the plasma membrane. Effects upon the activities of the cells must therefore be mediated by a receptor that relays information about the extracellular environment into the cell. In the case of steroid hormones, this extracellular/intracellular distinction is not maintained. It is generally accepted that when steroid hormones are present in the extracellular fluid, they may move through cell bodies and back into the extracellular space. It is only when the circulating steroid encounters an appropriate receptor complex that its actions affect cellular activity. The receptor complexes for steroids are usually found in the cytoplasm or in the nucleus of the

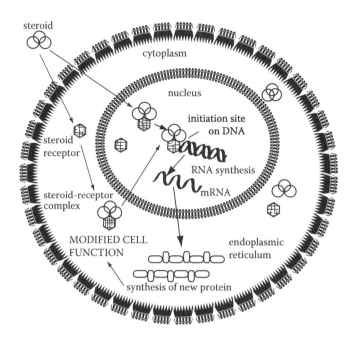

Figure 4.1 Classical view of steroid actions. Steroids cross the plasma membrane and bind to steroid receptors in the cytoplasm or the nucleus of the cell. The steroid-receptor complex attaches to the initiation site on the DNA, stimulating RNA synthesis and ultimately leading to the synthesis of new protein.

cell, where their primary action is to alter gene transcription. The receptors in the nucleus are associated with specific DNA sequences known as "promoters" (Figure 4.1). When the receptors in the cytoplasm or nucleus are activated by hormone binding, they form complexes that move into the nucleus of the cell and bind to the promoter sequences. In both cases, it is the activation of the promoter sequences that results in the synthesis of new protein by the cell. In the adult brain, protein synthesis can have a number of functions, including the production of new receptors or the synthesis of neurotransmitters or hormones. The effect on the ongoing function or activity of the cell may occur in minutes to hours, depending upon the amount of time required for protein synthesis to occur.

Methods for identifying receptors in the CNS

There are two prerequisites for the identification and localization of receptors. First, it is necessary to attach a marker molecule to the receptor so that the receptor can be visualized or counted in some way. Second, the labeled receptors need to be localized to specific brain regions.

In the first instance, the "grind and bind" methods, brain tissue is homogenized to allow maximum exposure of the receptors to the labeled drug. The homogenate is then centrifuged to separate the protein that contains the receptors and incubated with the radio labeled drug. Following incubation, the receptor-drug complexes are separated from the unbound ligand. The final ligand-receptor complex product can then be analyzed and characteristics such as B_{max} (an estimate of the number of binding sites) and K_D (a measure of the affinity of the receptor for the drug) can be determined. The main drawback of this type of analysis is that direct receptor localization is limited by the amount of tissue that can be dissected out for analysis and it gives no information about the spatial distribution of the receptors.

This second requirement, receptor localization, is best achieved using thin sections of brain tissue where receptors can be visualized directly. These sections of brain tissue are mounted on glass slides and incubated in medium containing the labeled ligand. Following incubation, the extra medium is removed and glass coverslips are placed on the slides to protect the tissue. The slides are then analyzed visually or by computer. The receptor sites can be localized in fine detail. The drawback is that the binding characteristics of the receptor are difficult to determine. Ideally, the two methods are used together to give detailed localization of the receptors as well as their binding characteristics.

Estrogen receptors

Two types of estrogen receptors (ERs) have now been cloned and sequenced. The first ER, which was cloned from rat uterus by Koike et al. in 1987, is a member of a much larger family of receptors, "ligand-dependent nuclear transcription factors" (Koike et al. 1987; Simerly et al. 1990). This family includes other steroid receptors as well as receptors for Vitamin D, thyroid hormone, and retinol. All of the receptors in this family have an active region, the DNA-binding domain, which enables the receptor to bind to the "promoter" sequence on the DNA. The DNA promoter sequence, which recognizes the ER binding domain, has been shown to be present in DNA in a number of brain regions. It has also been demonstrated that this binding domain is identical in rat and human (Koike et al. 1987). The cloning of the receptor was a significant step forward for estrogen researchers, because it meant that for the first time, there was a highly specific target for studying estrogen binding sites in the brain. Instead of looking for the uptake of radioactively labeled estrogen in brain cells, and then assuming the existence of receptors in those locations, it was possible to design probes that would bind specifically to the amino acid sequence of the estrogen receptor. Using these histochemical methods it soon became clear that the ERs identified by Koike et al.

could not account for all the estrogen binding observed in the brain. In 1996 Kuiper et al. cloned a second type of ER. The new receptor, which was found in rat prostate and ovary, was named ERβ making the earlier identified receptor ERα.

The two ER types have subsequently been identified in a number of brain regions. There is a general consistency in the literature on the distribution of the two types of ERs (Table 4.1). Both types of receptor are predominantly distributed in areas associated with the limbic system: the amygdala, septum, and hypothalamus. ER distribution is generally sparse in other regions such as the brain stem and cerebellum, although there is evidence for moderate numbers of ERα in selected brain stem nuclei, including the periaqueductal gray, locus coeruleus, and area postrema. A recent study (Laflamme et al. 1998) characterized ERα and ERβ in intact female and male rats. Laflamme et al. reported that overall there was a greater expression and wider distribution of ERαs than of ERβs. Interestingly, there was little difference in the number or distribution of receptors between female and male rats. Little is known about the function of the two types of ERs. However, it has been demonstrated that 17β-estradiol causes rapid and sustained activation of the microtubule-associated protein (MAP) kinase signaling pathway, a second messenger pathway activated by membrane-bound receptors (Toran-Allerand et al. 1999). Because the action on the MAP kinase pathway occurs in ERα knockout mice (see below), it has been suggested that this action is mediated by other estrogen receptors, such as the ERβ receptor.

The genomic actions of ERs have been characterized in a number of species including rat, mouse, monkey, and human. The primary genomic action of estrogen binding to the ER is to alter gene expression via action on DNA. There is, however, evidence that in some cases, estrogen can alter the turnover of mRNA (Simerly et al. 1990). Functionally, little is known about the

Table 4.1 Characteristics of two types of estrogen receptors.

ERα	ERβ
Distribution:	Distribution:
predominantly in amygdala, septum, hypothalamus also in periaqueductal gray, locus coeruleus, area postrema	predominantly in amygdala, septum, hypothalamus, olfactory cortex
greater expression than ERβ	smaller expression than ERα
number and distribution: F = M	number and distribution: F = M
genomic action	genomic action
membrane-bound receptors	membrane-bound receptors may act on MAP kinase pathway

ER's actions. The presence of receptors in areas associated with reproduction is to be expected, although even in those regions their actions are not understood. The presence of ERs in other brain regions provides a foundation for estrogenic modulation of a number of nonreproductive functions.

Steroid binding sites have also been observed on the extracellular membrane (Ramirez et al. 1996) and on classical neurotransmitter receptor complexes, suggesting that, in addition to their longer-term effects on protein synthesis, steroids are also able to produce or modulate rapid changes in synaptic activity. Membrane-bound receptors for estrogen, progesterone, and testosterone have been identified and characterized. Evidence suggests that the action of one type of membrane-bound estrogen receptor is G-protein coupled (Mermelstein et al. 1996). It has also been demonstrated that one of the membrane-bound estrogen receptors corresponds to a subunit of proton ATP synthase (see below).

Zheng and Ramirez (1999) have suggested another possibility for rapid-acting, nongenomic estrogen receptors. Using a preparation from rat brain that consisted of mitochondrial material, the authors isolated a protein that bound 17β-estradiol, the ligand for the ERs discussed above. Using the sequencing methods described in Chapter 1, the authors were able to identify the protein as a subunit of the mitochondrial enzymes ATP synthase/ATPase, which are essential for the production of adenosine triphosphate (ATP). ATP is the basic unit of energy, produced by the mitochondria and essential for biological function throughout the body. If the binding of estrogen to this protein can alter the energy metabolism of the cells, this provides a pathway for rapid changes in brain activity associated with estrogen.

Progesterone receptors

There are two types of genomic progesterone receptors, P-A and P-B. One of the most interesting properties of progesterone in the CNS, however, is its relationship to the $GABA_A$ receptor. It has been demonstrated that the rapid CNS effects of progesterone are due to its two neuroactive metabolites, allopregnanolone and pregnanolone. Both of these metabolites bind to the steroid binding site on the $GABA_A$ receptor, increasing the inhibitory activity of GABA. This is consistent with reported inhibitory effects of progesterone, including sedation and even anesthesia, but inconsistent with other reports of progesterone-induced depression and anxiety.

Nitric oxide is an elusive gas synthesized in the brain in response to stimulation by glutamate. It diffuses through cell membranes and acts on intracellular processes, usually causing excitatory effects. Nitric oxide has been identified as an important neurochemical in learning and memory processes. In the brains of healthy females, nitric oxide levels are negatively correlated with progesterone levels, and are lowest on cycle day 21, when progesterone peaks (Giusti et al. 2002).

"The knockout model"

A novel way of looking at the functional importance of specific receptor types is to look at the effects of genetically modifying an animal so that it does not express the receptor. These animals, usually mice, are known as "knockout" animals. In 1999, Rissman and colleagues reported interesting behavioral anomalies in male and female ERα knockout (ERαKO) mice. Given the importance of estrogen and its receptors in development, it probably came as something of a surprise to many researchers that ERαKO mice could survive the gestation period and be born alive. But survive they did, and they matured to adulthood looking very much like their wild-type (WT; not genetically modified) littermates. Probably the most obvious, and expected, differences in the adult ERαKO mice were in their sexual behavior. Neither the adult females nor the adult males displayed normal sexual behavior, and both females and males were infertile. Perhaps the most interesting differences, at least for our purposes, were the differences between the ERαKO and their WT littermates in the ability to learn spatial orientation tasks.

The Morris water maze is a tool long used by experimental psychologists to examine an animal's ability to learn tasks that depend upon the animal's memory for spatial orientation (Figure 4.2). The water maze itself is a pool with a movable platform located just below the surface of the water. Often powdered milk is added to the water to make it opaque and ensure that the animal cannot see the platform. The animal is trained over a number of trials to swim to the platform and wait to be removed from the pool (escape). Most animals learn the task easily, and the latency to reach the platform decreases over trials. The ERαKO mice that Rissman et al. (1999) tested learned to escape from the water maze just as well as their WT littermates. Both types of female mice were treated with high doses of estrogen. The ERαKO females showed the same pattern of learning as the untreated females, and their latency to find the platform decreased over trials. However, when WT females were treated with estrogen, they failed to learn the task, and the latency to find the platform and escape did not decrease with practice. In this case, it appeared that high levels of estrogen, acting via the ERα receptor, disrupted the learning process. We will return to the question of estrogen and learning in Chapter 6.

In addition to the action of estrogen alone on CNS neurons, the ability of estrogen to interact with neurotransmitter systems is now well documented. The neurotransmitters to receive the greatest attention in terms of estrogen interactions are GABA, serotonin, and dopamine. Interestingly, these three neurotransmitter systems are implicated in the etiology of

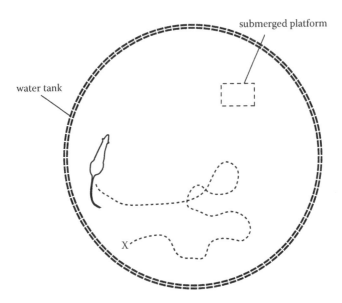

Figure 4.2 Schematic representation of a water-maze. The rat is placed in the maze at point "x" and must swim to the submerged platform to escape. Over repeated trials the rat learns to swim directly to the platform. The time taken to reach the platform and/or the distance traveled may be measured.

anxiety, depression, and psychosis, disorders where sex differences have been well established.

GABA

Gamma-aminobutyric acid (GABA) is an amino acid neurotransmitter with inhibitory actions. GABA receptors are widely distributed through-out the brain. The GABA receptor is a transmembrane receptor complex either with an associated ion channel or a G-protein. The receptors are composed of a number of basic subunits that may be arranged in different configurations to produce different receptor subtypes with specific binding characteristics. There are three main subtypes of GABA receptors, $GABA_A$, $GABA_B$, and $GABA_C$ (Table 4.2). In addition, there are numerous subtypes of $GABA_A$ receptors, approximately 14, which are made up of different combinations of the subunits. Binding of GABA to the GABA site on the $GABA_A$ receptor complex causes an influx of chloride ions through the associated ion channel. When this occurs, the probability that the cell will produce action potentials decreases and the overall effect is a temporary suppression of cellular activity. Binding of GABA to the $GABA_B$ receptor complex causes an influx of potassium ions through the associated ion channel, which also results in inhibition.

Table 4.2 Characteristics of GABA receptors.

Receptor subtype	Characteristics
GABA$_A$	Ligand-gated ion channel
	Chloride influx
	Post-synaptic inhibition
GABA$_B$	G-protein-coupled
	Inhibition of adenylate cyclase
	Pre-synaptic inhibition, decreased calcium
	Post-synaptic inhibition, increased potassium influx
GABA$_C$	Ligand-gated ion channel
	Chloride influx

It is the GABA$_A$ receptor that is the target of drugs that depress CNS activity, such as alcohol and sedatives. In fact, the GABA$_A$ receptor complex contains binding sites for alcohol, barbiturates, benzodiazepines, *and* steroids (Figure 4.3). The preoptic area of the limbic system is rich in receptors for estrogen, progesterone, and testosterone. It also contains large numbers of GABA-containing neurons. The first documentation of an interaction between estrogen and GABA in the CNS was in 1983. Sar et al. (1983) reported that radioactively labeled estrogen was taken up by GABA-containing neurons in the preoptic region. Since that time, the interactions between estrogen and GABA have been widely studied, using more and more sophisticated techniques. It is now known that about 20 percent of neurons in this region that contain GABA (as measured by glutamic acid decarboxylase [GAD] reactivity) also contain estrogen receptors. Administration of estrogen has been demonstrated to increase the extracellular concentration of GABA, but not apparently through the increased synthesis of GABA by GAD. Herbison (1997) has reviewed evidence that suggests that the increased extracellular concentration of GABA in the preoptic region may be due to an estrogen-stimulated increase in the activity of GAT-1, the GABA transporter. This is not necessarily the case for all GABA-estrogen interactions in the brain. For example, Weiland (1992) has demonstrated that in the hippocampus, both estrogen and progesterone modulate GAD activity, and, therefore, the concentrations of GABA. Interestingly, it has been demonstrated in the preoptic area and in the ventromedial nucleus of the hypothalamus that the level of activity of GABA-containing neurons in male rats is twice that of female rats (Grattan and Selmanoff 1997)

The GABA$_A$ receptor has been demonstrated to be a target for direct actions by hormones via the steroid binding site (see Majewska 1991 for a review) (Figure 4.3). The naturally occurring steroids that bind to this site include androsterone, tetrahydrodeoxycorticosterone, and the

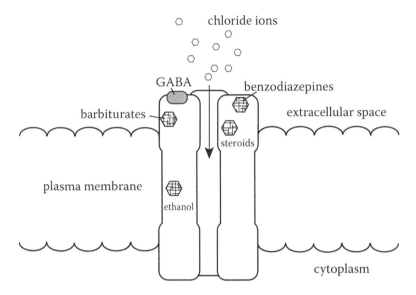

Figure 4.3 Schematic representation of the GABA$_A$ receptor. In addition to the GABA binding site, the receptor also contains allosteric binding sites for barbiturates, benzodiazepines, ethanol, and steroids.

progesterone metabolite, tetrahydroprogesterone. It has been demonstrated in a number of studies, using biochemical and electrophysiological techniques, that the binding of these steroids to the allosteric binding site on the GABA$_A$ receptor prolongs the neural response to GABA. For example, steroid binding has been demonstrated to increase the duration of the Cl$^-$ influx through the GABA$_A$ receptor ion channel and to prolong GABA-mediated inhibitory postsynaptic potentials (Majewska 1991). This is particularly important because results of this type clearly distinguish a membrane receptor action from a genomic effect.

It has also been demonstrated that two anabolic steroids, stanozolol and 17α-methyltestosterone, when bound to the GABA$_A$ receptor, modulate the binding of benzodiazepines to the benzodiazepine binding site on the GABA$_A$ receptor, and that the effects differ between females and males (Masonis and McCarthy 1995). This result will be further discussed in Chapter 8.

Serotonin

In contrast to the widespread distribution of GABA in the brain, serotonin (5-hydroxytryptamine, 5-HT)-containing neurons are found in discrete regions in the midbrain and brain stem. All of the brain's serotonergic projections arise from these two areas (Figure 4.4). There are at least 16 subtypes of serotonin receptor, and the effect of serotonin will depend

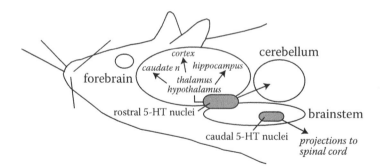

Figure 4.4 All 5-HT projections arise from two 5-HT neurone-containing regions (shown in gray) in the midbrain and brain stem, respectively.

upon the receptor subtype to which it binds. The effect may be inhibitory or excitatory and may involve increases or decreases in cAMP, increased potassium conductance, or activation of phosphoinositide-related second messenger pathways. Although it is a gross simplification, serotonin is often said to be the neurotransmitter associated with depression, migraine, and the action of hallucinogenic drugs such as "magic mushrooms" and LSD. In fact, serotonin is involved in a wide range of neural functions including sleep, pain, and various endocrine actions. Probably one of the most publicized actions of serotonin is in the mediation of depression. Fluoxetine (Prozac), paroxetine (Aropax), and other drugs in the selective serotonin reuptake inhibitor (SSRI) family act by increasing the time that serotonin (as well as noradrenaline and, to some extent, dopamine) remains in the synapse.

There is a long (but not entirely consistent) history of studies showing sex differences in serotonin levels and actions, dating from the 1950s (Table 4.3). From a number of animal studies, it appears that during times when estrogen and progesterone are naturally high, serotonergic activity is decreased. In ovariectomized female rats, where the effect of estrogen application may be more directly observed, both serotonin levels and receptor binding are modulated by estrogen. On first observation, the results of the various studies appear inconsistent and confusing. When broken down by brain region and receptor type, a pattern begins to emerge, however. There is an increase in extracellular serotonin in the suprachiasmatic nucleus, the median eminence, and the dorsal raphe (a region containing serotonergic neurons) following estrogen injection but a decrease in the cortex, medial preoptic nucleus, and anterior hypothalamus. When the results of the receptor binding studies are organized in a similar manner, binding to 5-HT_1 receptors appears to decrease in the cortex but increases in the midbrain, and in the majority of reports, in the hypothalamus. Binding of serotonin to 5-HT_2 receptors increases in the

Table 4.3 5-HT and estrogen interactions.

5-HT$_1$ receptor binding (rat)	Decreased with high estrogen
extracellular 5-HT (OVX rat)	Increased by estrogen in: suprachiasmatic nucleus, median eminence, dorsal raphe
	Decreased by estrogen in: cortex, medial preoptic nucleus, anterior hypothalamus
5-HT$_1$ receptor binding (OVX rat)	Increased by estrogen in: midbrain, hypothalamus
	Decreased by estrogen in cortex
5-HT$_2$ receptor binding (OVX rat)	Increased by estrogen in: cortex, nucleus accumbens, dorsal raphe
5-HT synthesis (human)	m > f

cortex, nucleus accumbens, and dorsal raphe. Rubinow et al. (1998) published a comprehensive review of the literature on interactions between estrogen and serotonin, which covers both human and animal studies. For readers who would like to pursue this area further, this review is an excellent place to start. There is little direct evidence available on serotonin levels in the human. However, a PET imaging study of healthy females and males has shown that in many areas of the brain, females synthesize less serotonin than their male counterparts. Overall, the rate of synthesis was found to be 52 percent higher in males (Nishizawa et al. 1997).

Dopamine

Dopamine, like serotonin, is found only in neurons in discrete brain regions. The dopaminergic pathways, which arise from these regions, are anatomically and functionally distinct (Figure 4.5). The nigrostriatal pathway projects from the substantia nigra to the striatum. This pathway contains about 75 percent of the brain's dopamine. Loss of dopaminergic neurons in the substantia nigra occurs in Parkinson's disease, depriving the striatum of its dopaminergic input. The mesolimbic and mesocortical pathways project from the dopamine-containing neurons in the ventral tegmental area to parts of the limbic system and the neocortex. Excessive dopamine activity in this pathway is associated with the development of psychosis, and modulation of the activity of the dopamine receptors is one of the chief strategies for the treatment of schizophrenia. The third dopamine pathway, the tuberoinfundibular pathway, projects from the arcuate nucleus of the hypothalamus to the pituitary gland. This pathway plays a central role in the modulation of prolactin release. As with the other neurotransmitters discussed so far, there are a number of dopamine receptor subtypes. The D$_1$, D$_2$, and D$_4$ receptors are the subtypes primarily associated with the treatment of psychosis.

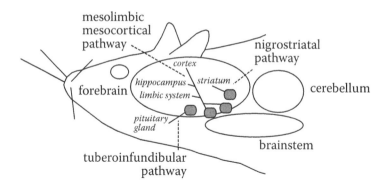

mesolimbic
mesocortical
pathway

nigrostriatal
pathway

cortex

forebrain

hippocampus *striatum*
limbic system

cerebellum

pituitary
gland

brainstem

tuberoinfundibular
pathway

Figure 4.5 The three dopaminergic pathways the nigrostriatal pathway arising from the substantia nigra *pars compacta*, the mesolimbic/mesocortical pathways arising in the ventral tegmental area, and the tuberoinfundibular pathway arising from the hypothalamus. Gray areas represent the regions of dopamine containing neurons.

Dopamine has a well-established role in female reproductive functions. Via the tuberoinfundibular pathway, dopamine released from neurons in the hypothalamus acts on dopamine receptors on lactotrophs (prolactin-secreting cells) in the anterior pituitary to modulate prolactin release. In this sense, dopamine may be classed as a hormone as well as a neurotransmitter. The effect of dopamine on lactotrophs is to reduce prolactin secretion. Administration of estrogen over a long period induces increased prolactin secretion and a condition known as hyperprolactinaemia. It has been suggested that estrogen may reduce the sensitivity of the dopamine receptors on the lactotrophs, resulting in decreased inhibition in response to dopamine binding and therefore, increased prolactin secretion (see Johnson and Everitt 1995 for a review).

The striatum is an area crucial for motor control. Dopamine released from neurons in the substantia nigra pars compacta inhibits GABA-containing neurons in the striatum, which project to the cortex by way of the thalamus. The nigro-striatal pathway forms the foundation for a complex and essential network for motor control. The importance of this system is clearly demonstrated when disease causes a disruption of the network. Both Parkinson's disease (where the dopamine is lost) and Huntington's disease (where GABA is lost) are characterized by profound and irreversible disruptions of motor control.

There is a good deal of experimental evidence demonstrating that estrogen affects striatal dopaminergic activity. For example, using the OVX rat model, it has been demonstrated that a single injection of 17-β estradiol (but not progesterone) increased the binding of dopamine in the striatum by 24 percent while leaving the affinity unchanged (Morissette et al. 1990a) (Table 4.4).

It has also been demonstrated that the density of striatal dopamine receptors changes during the estrus cycle in female rats. The density of D_1 receptors has been demonstrated to be greatest on diestrous II. In OVX rats, the density of D_1 receptors decreased by 17 percent. It is interesting in these two studies that the variable that changes is the actual density or number of sites, but not the affinity of the dopamine receptor (Levesque et al. 1989). By contrast, the affinity of the D_2 receptor has been reported to change within 15 minutes of 17β-estradiol administration (Levesque and Di Paolo 1988). It has also been demonstrated that the D_2 receptor affinity fluctuates during the rat estrus cycle (Di Paolo et al. 1988) (Table 4.4).

Not only has estrogen been demonstrated to alter the characteristics of dopamine receptor binding, but it has also been reported to alter dopamine metabolism in the striatum. A study using the OVX rat model has demonstrated that a single injection of 17β-estradiol, progesterone, or a combination of the two, caused an increase in dopamine metabolites (including homovanillic acid) that peaked 15 to 60 minutes after the injection. However, the concentration of dopamine itself increased only following progesterone or progesterone combined with estradiol (Morissette et al. 1990b) (Table 4.5).

Homovanillic acid (HVA) in plasma or urine is commonly used as a measure of brain dopamine metabolism. A study of HVA at different stages of the menstrual cycle in healthy women failed to find any significant changes (Abel et al. 1996). However, a study of HVA levels in the CSF of OVX monkeys and post-hysterectomy women showed significant increases in HVA (Di Paolo et al. 1989).

The acute administration of amphetamine to rats or mice causes a dose-dependent increase in the release of dopamine in the striatum. Because of the reliability of the effect, "amphetamine-stimulated striatal dopamine release" is an animal model frequently used to study the interactions of

Table 4.4 The effect of estrogen on dopamine receptors.

Variable	Effect
Binding DA, OVX rat	Increased 24 percent in striatum by estrogen
Density of D_1 receptors, OVX rat	Decreased by 17 percent
Density of D_1 receptors, I/F rat	Greatest diestrous II (high progesterone/ low estrogen)
Affinity D_2 receptor, OVX rat	Altered within 15 min. of estrogen administration
Affinity D_2 receptor, I/F rat	Fluctuated during estrous cycle

DA, dopamine; I/F, intact female; OVX, ovariectomized female.

Table 4.5 Dopamine and the estrous/menstrual cycle.

Variable	Effect
Dopamine metabolism, OVX rat	HVA increased by estrogen, progesterone, estrogen+progesterone
Dopamine metabolism, human F	No cycle-related changes HVA
Dopamine metabolism, OVX monkey, hysterectomized human	Increased HVA
Amp stimulated DA release, I/F rat	Greatest estrus (low estrogen)
Amp stimulated DA release, OVX rat	Decreased, but restored by estrogen
DA transporter, OVX rat	Decreased striatum, nucleus accumbens, not restored by estrogen
DA transporter mRNA, OVX rat	No change after two weeks (in short-term animals)
	Increased after two weeks (in rats three months post-OVX)
	Modulated by estrogen, substantia nigra
	No effect estrogen, ventral tegmental area
DA transporter availability, human, SPECT	Not related to sex or cycle
TH immunoreactivity, OVX monkey	Decreased dorsolateral prefrontal cortex, restored by HRT.
DA β-hydroxylase, OVX monkey	Increased dorsolateral prefrontal cortex, modulated by HRT.

Amp, amphetamine; DA, dopamine; I/F, intact female; HRT, hormone replacement therapy; HVA, homovanillic acid, a DA metabolite; OVX, ovariectomized female; TH, tyrosine hydroxylase.

drugs and other neurotransmitters with dopamine. A number of studies have demonstrated that amphetamine-stimulated dopamine release can be modulated by administration of estrogen or progesterone. Female rats exhibit the greatest response to amphetamine on the day of estrus. This increased response includes greater striatal dopamine release, metabolism, and concentration compared to other days of the cycle. In OVX rats, the response to amphetamine is greatly decreased but returns with estrogen replacement. Results of this kind suggest a role for estrogen in modulating dopamine release, but give no insight into possible mechanisms of action.

One area of interest has been the dopamine transporter (DAT). Following release, dopamine is removed from the synaptic cleft by binding to the DAT and taken up into the presynaptic terminals. A study of the effects of OVX on DAT (Bossé et al. 1997) has yielded particularly interesting results. Three groups of rats were used in the study: long-term OVX rats that were tested three months after OVX; short-term OVX, that were tested after only two weeks; and control rats that received no operation.

Half of the animals in each group received 17β-estradiol for two weeks. The other half received vehicle injections. The two-week treatment protocol meant that the short-term animals were treated from the time of OVX while the long-term animals were treated for the last two weeks of the three-month period. At the end of the experimental time, the animals' brains were removed and prepared for analysis. DAT mRNA expression in the substantia nigra *pars compacta* and the ventral tegmental area was measured using in situ hybridization. DAT levels in the striatum and nucleus accumbens were measured using autoradiography.

The authors reported that DAT levels decreased in the striatum and nucleus accumbens following OVX and that treatment with 17β-estradiol did not restore it (Table 4.5). The expression of DAT mRNA was unchanged in the short-term animals but was increased in the long-term animals. In the long-term animals, 17β-estradiol treatment partially restored the mRNA levels in the substantia nigra but had no effect in the ventral tegmental area. This latter result is particularly interesting because it suggests that the effect of estrogen replacement may differ between the dopamine systems in humans. While motor control systems (the nigrostriatal pathway originating in the substantia nigra *pars compacta*) may benefit from hormone replacement, the mesocortical/mesolimbic systems may be more resistant to treatment. Contrary to the reports from animal studies, a SPECT analysis of the striatal dopamine transporter in humans failed to find a difference due either to sex or phase of the menstrual cycle (Best et al. 2005).

The mesolimbic dopamine pathway has often been associated with the etiology of psychosis. Hyperactivity in this pathway, resulting in excessive dopamine release, is thought to be associated with the development of cognitive/affective disorders (Goldstein and Deutch 1992). Estrogen has long been thought to play a protective role, but there has been little direct evidence to support this belief. Recently, however, Kritzer and Kohama (1998) demonstrated that in adult female rhesus monkeys, OVX decreased immunoreactivity for tyrosine hydroxylase in the dorsolateral prefrontal cortex (which would effectively decrease dopamine levels). The tyrosine hydroxylase levels could be restored by hormone replacement. In addition, using immunocytochemistry for dopamine β-hydroxylase, choline acetyltransferase, and serotonin, the levels were also altered by OVX. However, this is the really interesting part. While there was a small decrease in choline acetyltransferase, there was a pronounced increase in immunoreactivity for dopamine β-hydroxylase and serotonin. When the authors looked at the effects of hormone replacement, they found that estrogen alone and estrogen combined with progesterone normalized choline acetyltransferase and dopamine β-hydroxylase activity. It was the estrogen plus progesterone therapy that was most effective in normalizing serotonin, however. In their discussion, the authors propose that estrogen and progesterone may be important modulators of neurotransmitter actions in the prefrontal cortex. Another interesting aspect of dopamine function is its role in neurotoxicity. Six-hydroxydopamine (6-OHDA) has frequently been used to lesion the striatum in an effort to produce an animal model of Parkinson's disease. When 6-OHDA lesions are produced in OVX rats, estrogen treatment has been demonstrated to reduce the extent of the lesions compared to vehicle-treated controls. Similarly, estrogen replacement has also been demonstrated to be protective against MPTP-induced neurotoxicity. We will return to the discussion of estrogen as a neuroprotectant in Chapter 8.

Cannabinoids

Cannabinoids are a group of drugs derived from the active ingredients in *cannabis sativa*, including Δ^9-tetrahydrocannabinol (Δ^9-THC), cannabinol, and cannabidiol. Although cannabis has been used for thousands of years to relieve pain, the receptors for cannabinoids and endogenous cannabinoids were only discovered in the late 1980s/1990s. There are at least five endogenous cannabinoids (endocannabinoids), all derived from arachadonic acid. The first endocannabinoid to be discovered, anandamide, is named after the Sanskrit word "ananda," meaning "bliss." There are two known types of cannabinoid receptors, CB_1 and CB_2, which are found throughout the body. Initially, CB_1 receptors were thought to be restricted to the brain while CB_2 receptors were thought to be found in other parts of the body. Now it is known that both receptor types are found throughout the body, with CB_1 primarily in the brain and CB_2 in greater number in areas such as cardiovascular and immune systems. Cannabinoid receptors are coupled to inhibitory G-proteins and are unusual because they are located pre-synaptically and are activated by a "retrograde transmitter," anandamide (Figure 4.6). The endocannabinoids that activate the CB_1 and CB_2 receptors are synthesized, as needed, in the post-synaptic neurons. Endocannabinoids are released into the synapse and cross to the pre-synaptic side where they bind to their receptors and inhibit presynaptic release of GABA, glutamate, and other neurotransmitters (Benarroch 2007).

Sex differences have been reported in the behavioral effects of cannabinoids and also in their metabolism. In a study using rats, the animals were injected with Δ^9-THC, and the amounts of the drug were analyzed in blood and brain tissue. The levels of Δ^9-THC were found to be similar in the blood and brain of females and males; however, the metabolites of Δ^9-THC were significantly higher in the brains of the female rats. The authors conclude that the higher levels of Δ^9-THC metabolites in the brains

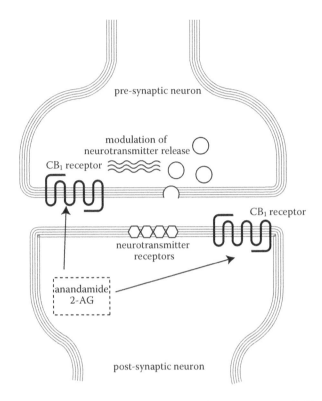

Figure 4.6 Schematic representation of pre- and post-synaptic of CB$_1$ receptors, 2-AG, sn-2-arachadoylglycerol.

of the females was consistent with the more intense effects of Δ^9-THC on behavior observed in female rats (Tseng et al. 2004).

A study of CB$_1$ receptor binding using positron emission tomography (PET; see next section) in healthy adults demonstrated a similar distribution of the receptors in females and males. In young adult females there was a significantly lower amount of the tracer remaining intact in their blood plasma, consistent with higher metabolic rate for the cannabinoid, resulting in greater availability. In females but not males, CB$_1$ receptor binding increased with increasing age in several brain regions, including the basal ganglia, temporal cortex, and limbic system with particularly prominent changes noted in the hippocampus (Van Laere 2007). There were no age-related decreases in any region.

Measures of global function

As the techniques for measuring global function in the living brain become more refined, the information value of the methods becomes greater and greater. While immunohistochemistry can identify specific neurotransmitters in individual axons, an MRI provides a measure of neuronal activity for a brain region or regions. The responses of individual neurons are currently too small to be seen individually. They sum to produce images representing activity in a region. With the current rate of technical advances, however, it shouldn't be long before imaging at the level of the individual neuron becomes a possibility. It is already clear that imaging of the living brain may produce evidence on neurotransmitters and receptors that is very different from the information obtained from surgically excised tissue or by postmortem tissue analysis. Already imaging studies have rewritten large parts of the texts on neuroanatomy and neurophysiology. A 2007 review of the information on sex differences in structure and function resulting from imaging studies is essential reading for anyone interested in the subject (Cosgrove et al. 2007).

There are three types of global measures to consider (Table 4.6). The visual imaging techniques, positron emission tomography (PET) and functional MRI, provide anatomical localization of activity. Electroencephalography (EEG) provides activity patterns but with only gross anatomical localization referenced to scalp surface electrodes.

EEG measures neural activity, "brain waves," by recording the patterns of electrical activity from an array of electrodes attached to a person's scalp. The placement of the electrodes is standardized to skull landmarks using an international protocol known as the "10–20 System." Each electrode in the 21-electrode array is numbered, odd numbers on the left, even numbers on the right (Figure 4.7). A letter preceding the number indicates the scalp region. So, O1 is the electrode on the left side of the occipital region, while T4 refers to an electrode over the right temporal region. Adoption of the 10–20 International System has allowed a standardization of EEG recording between individuals and in a single individual across time (Neidermeyer and Lopes da Silva 1993). The activity

Table 4.6 Methods of measuring global brain activity.

Method	Information
EEG	Ongoing, event-related, electrical activity
	Cortical activity only
	Gross signal localization
PET	Functional brain image
	Measures regional uptake of radio-labeled isotope
	Resolution, 4–8 mm
Functional MRI	Functional brain image
	Sensitive to changes in tissue structure
	High resolution, 0.1 mm

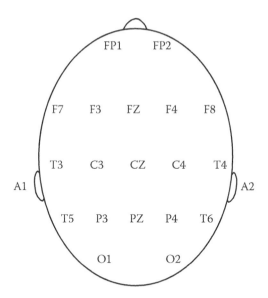

Figure **4.7** Standard electrode array for EEG recording. Odd number indicates left side; even number indicates right side. A, auricle; C, central; CZ, vertex; F, frontal; FP, frontal pole; FZ, frontal midline; O, occipital; P, parietal; PZ, parietal midline; T, temporal. From Niedermeyer and Lopes da Silva 1993.

measured from any given electrode is a sum of the activity in the region of cortex below the electrode. Specifically, it is thought to be the summation of post-synaptic potentials in the area recorded by the electrode. While EEG cannot give information about the ongoing activity in deep brain structures, it provides a valuable insight into the patterns of cortical activity. It is a valuable diagnostic tool, particularly in the case of epilepsy. It is also a research tool for studying the changes in cortical activity associated with the performance of different tasks or the administration of drugs.

When recording from normal individuals, the EEG will be a complex wave form, made up of individual frequencies of potentials ranging from 0.1 to 70 Hz. Frequency analysis of the EEG component wave forms usually reveals four frequency components that are characteristic of cortical activity (Table 4.7). The wave form associated with wakefulness, alpha waves, has a frequency range of 8–13 Hz. Alpha waves are recorded over the posterior regions of the brain. They are usually recorded while the person relaxes with her eyes closed, hence the term "relaxed wakefulness." Alpha waves can be temporarily blocked by an arousing stimulus, for example, by opening the eyes, or by performing a task such as mental arithmetic. Recordings of alpha waves are generally agreed to be asymmetrical, with the highest voltages recorded over the nondominant hemisphere. Beta waves range from 13 to 35 Hz and are associated

Table 4.7 Characteristics of EEG frequencies in humans.

Frequency	Characteristics
Alpha	8–13 Hz
	Relaxed wakefulness
	Highest over nondominant hemisphere
Beta	13–35 Hz
	High-level mental activity
	Frontal and central regions
Delta	< 3.5 Hz
	Sleep
Theta	4–7 Hz
	Sleep, drowsiness

Source: Niedermeyer and Lopes da Silva (1993).

with high levels of activity, such as performing difficult mental arithmetic. Beta waves are recorded from the frontal and central regions. Beta recordings from the central electrodes can be blocked by motor activity or tactile stimulation. Some drugs, such as barbiturates, have been demonstrated to increase both the quantity and voltage of beta activity. Delta waves (<3.5 Hz) and theta waves (4–7 Hz) are associated with sleep. Theta, named for its origins in the thalamus, is important in infancy and childhood, but in adults is primarily associated with sleep. A normal, awake adult EEG will show only a small amount of activity on the theta band and no organized "theta rhythm." This is in marked contrast to the theta rhythm recorded in rodents, which is recorded during alertness and has been suggested to be a measure of "emotion."

Studies using EEG may measure and analyze several different variables for a single recording period. First, the EEG can be analyzed for the presence or absence of the different component wave forms. Initially, this kind of analysis was performed by passing the signal through an assortment of filters to physically remove different frequencies of activity. Now, computer analysis allows online Fourier analysis of the EEG into the component frequencies. Further analysis may be used to determine the amplitude of the wave form (in mV), the total amount of time in which the wave form occurs in the recording period, and the timing of the wave form recorded from different electrodes (in-phase or out-of-phase).

The usual condition for measuring EEG, lying down or sitting quietly with the eyes closed, is assumed to measure EEG activity when the brain is alert but resting. It is impossible to know exactly what the brain is doing during that "resting" time, however. Most people find it nearly impossible to think of "nothing" while remaining alert. At least some portion of the variability in resting EEGs is probably attributable to the thought processes of the subjects. One way of adding an element of control to the EEG recording is to record the EEG following the presentation of a specific stimulus. Evoked potentials are EEG recordings of activity following a specific event. For example, auditory-evoked potentials may be recorded following a "click" stimulus delivered to the ear through headphones. When the activity in the period immediately following each click is recorded and averaged over a number of clicks, a distinct pattern of activity emerges. The resulting pattern will consist of a series of positive and negative components (peaks) that represent different stages of processing in the auditory pathway (Figure 4.8). An interesting aspect of evoked potentials is that the amplitude of the potential that is recorded is correlated with the size of the person. For this reason, a necessary step in the analysis of evoked potentials is to normalize the data with respect to body size. Many of the early evoked potential studies did not include this form of correction and so the data must be considered cautiously.

PET and functional MRI differ from CT by providing images of brain function at a given point in time. PET is similar to CT in its use of sequential, rotating images. But it differs in the sense that the signals being detected by the scanning equipment are of the distribution of positron-emitting (radioactive) isotopes in the brain tissue (resolution 4–8 mm). Positron-emitting isotopes can be bound to a number of biologically active substances to give a wide range of possibilities for functional analysis. For example, by binding an isotope to 2-deoxyglucose, the metabolism of glucose (an indicator of neuronal activity) can be measured. In addition to labeling glucose, neurotransmitters and drugs can also be labeled and imaged. Functional MRI uses the same equipment and methods as MRI

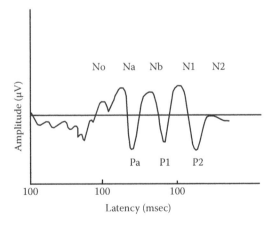

Figure 4.8 Schematic representation of an auditory-evoked potential. Signals are averaged over a number of trials to produce the characteristic wave forms.

scanning used for anatomical localization. The difference with functional MRI is that the scans are taken while the subject performs a specific task, for example, mental arithmetic. The question with functional MRI is "How does the image change when the subject performs the task instead of resting quietly?"

Evidence from EEG studies

One of the earliest reports of sex differences in EEG recordings was made at the 1967 meeting of the Central Association of Electroencephalographers. Giannitrapani and Snekhaus (Small 1967) reported that while conducting an EEG study of laterality they noted that in some brain areas differences in activity seemed to be associated with the sex of the subject. As a result of this observation, the authors conducted a study looking specifically at sex differences and the relationship between the activity of specific EEG electrode pairs. The EEGs of 15 males and 15 females were recorded and the authors reported that in-phase activity was decreased in male subjects during a psychomotor task, but was unchanged in females. The cortical areas in which the differences occurred were the left compared to the right prefrontal areas, the right prefrontal area compared to the right frontal areas, the right frontal area compared to the right motor areas, and the right motor area compared to the right parietal areas.

Since 1967 it has been demonstrated in a number of studies that there are sex differences in EEG activity that can be seen in childhood and that continue into old age. For example, Matsuura et al. (1985) analyzed the alpha, theta, beta, and delta wave activity of 1,416 people ranging in age from 6 years to 39 years. Analysis was made for three regions: frontal, central, and occipital. The greatest sex differences were for beta waves, for which the percentage of time during which beta could be recorded was greater for females for all three regions at all ages. The amplitude of the beta waves was also greater in females than in males after the age of 21 years. Alpha waves recorded in the occipital region were of greater amplitude in females compared with males after the age of 21, but the percentage of alpha time in the occipital region was greater in males than in females after age 17. The percentage of time for theta activity was higher in the frontal and central regions for females compared with males until age 25. No differences were reported for delta waves. The authors concluded that their results support the view that EEG matures faster in males than in females. However, they also suggested that because the differences in theta were small and no differences were found in delta, sex differences are not important in interpreting routine EEGs. This comment is a little surprising given that theta and delta are associated with sleep in adults and routine EEG recordings are made in awake, resting people.

In elderly people (aged 60–87 years), differences in central recordings of alpha, beta, and theta have been reported. The mean frequency of beta was reported to be increased in females, while alpha and theta frequency were decreased in females relative to males. The reported differences were specific to sex, independent of age. In their conclusions, the authors suggested that consideration of sex is an important issue in EEG analysis, at least for the elderly population (Brenner et al. 1995).

EEG changes that correlated with the different phases of the menstrual cycle have been reported by several researchers. The results of these studies, as summarized by Niedermeyer and Lopes da Silva (1993), are represented in Figure 4.9. The most consistently reported changes are in the frequency and percentage of alpha activity. At first glance, the pattern of EEG changes in relation to absolute hormone levels seems inconsistent. If one considers the EEG in relation to the female changes in hormone levels, however, a pattern begins to emerge. Alpha frequency is decreased during the menstrual phase, but increases when estrogen rises in the preovulatory phase. During the postovulatory phase, when estrogen decreases and then rises slightly, alpha frequency decreases.

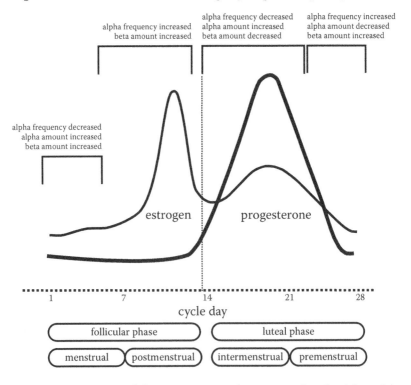

Figure 4.9 Changes in EEG frequency across the menstrual cycle. Adapted from Niedermeyer and Lopes da Silva 1993, p. 135.

Finally, when estrogen drops during the premenstrual phase, alpha frequency increases again. At the start of the next cycle, estrogen is still low and alpha frequency decreases until another change occurs, and the cycle repeats.

Evidence from evoked potential studies

The above studies measured the EEG in people resting with their eyes closed. Event-related potentials measure EEG activity associated with the occurrence of a particular stimulus, for example, a flash of light (visual-evoked potentials) or a specific sound (auditory-evoked potentials).

Visual-evoked potentials stimulated by a flash of light have been reported by Kaneda et al. (1996) to have a greater peak-to-peak amplitude in females than in males. This study is particularly interesting for two reasons. First, a large number of subjects, 100 females and 100 males, was tested. Second, the data were normalized to the body size of the subjects. This is an important factor in evoked potential recordings that, at least in the past, has often been overlooked. Because people with smaller bodies usually have proportionally smaller brains, the evoked potentials have less distance to travel and the latency will appear to be shorter. Initially, female/male differences were observed on several of the measured variables. When the body size adjustment was applied, however, the only remaining difference was in peak-to-peak amplitude. In looking at their female subjects alone, after adjustment for body size, the peak-to-peak amplitude of the P5 to N7 component was significantly greater during the luteal phase of the cycle than in the follicular phase. Interestingly, differences in latency, both between females and males, and between the luteal and follicular phases in females, disappeared when the body size adjustment was made.

Other studies relating visual-evoked potentials to the menstrual cycle have yielded mixed results, possibly because adjustments for body size were not made during data analysis. The P300 wave has been reported to increase in amplitude with high progesterone, increase in latency at ovulation, decrease in latency at ovulation, or not change at all. Interestingly, a subsequent study by Kaneda et al. (1997) of 44 females, with data normalized for body size, reported no significant differences overall between the luteal and follicular phase for either latency or amplitude of any of the visual-evoked potential components. The majority of studies using auditory- or somatosensory-evoked potentials also report no differences in latency or amplitude related to the menstrual cycle. Latencies, however, have been reported to be significantly shorter for females than for males.

Evidence from scan studies

Several studies using PET have looked for differences in the levels or patterns of brain activity between females and males (Table 4.8). For example, a study measuring cerebral blood flow during a series of cognitive tasks (Espisto et al. 1996) reported that the rate of blood flow during the various tasks was higher in the female subjects than in the males. A study of glucose metabolism failed to find differences in overall glucose utilization (Azari et al. 1992). The authors did find differences in the patterns of interaction between cortical areas (a similar idea to looking for synchronous and non-synchronous EEG activity), however. The correlations in activity between different cortical areas were more positive for females than for males. The areas where correlations were more positive for females were the left frontal and sensorimotor areas, while the correlations were higher for males in the right sensorimotor and occipital areas. This study is interesting from the perspective of functional laterality (Chapter 6). Although there are differences in processing between the two cerebral hemispheres in males and females, the same amount of energy is still used. Another

Table 4.8 Summary of results from human studies of global function.

Blood flow during cognitive tasks	f > m
Cerebral blood flow	f > m
Global metabolism	f = m
5-HT synthesis	f < m
5-HT$_2$ receptor binding	Varying results possibly due to metabolism differences
5-HT$_2$ receptor binding, post-menopause	Cortical binding higher with HRT
5-HT transporter	f > m
Cortical mACh receptors	f > m
D$_2$ receptor binding affinity	f < m, L striatum
DA transporter	f > m striatum
DA synthesis	f > m
Cortical GABA	f > m
μ-opioid receptor affinity	Follicular = luteal
	-ve correlation with E at follicular phase, hypothalamus, amygdala
Functional MRI:	
Visual cortex activation	m > f; m, R > L; f, L = R

Source: Cosgrove et al. (2007). E, estrogen; L, left; R, right.

PET study found a significantly higher rate of metabolism in females in the cerebellum only (Volkow et al. 1997).

PET imaging has also been used to look for neurotransmitter differences between females and males. It has been demonstrated that the rate of serotonin synthesis is 52 percent greater in the brains of males than females and that the binding capacity of the $5HT_2$ subtype of serotonin receptor is greater in males in the frontal and cingulate cortices (Biver et al. 1996). The dopaminergic system has also been reported to differ, with D2 receptors in the striatum of females showing a lower affinity for the radioactively labeled ligand, raclopride, than striatal D_2 receptors in males (Pohjalainen et al. 1998).

A PET study of opioid receptors failed to show any differences in the affinity of μ-opioid receptors between the follicular and luteal phases (Smith et al. 1998). However, the study did show a significant negative correlation between plasma estrogen levels and μ-receptor binding in the hypothalamus and amygdala during the follicular phase.

A recent functional MRI study has revealed particularly interesting characteristics of the action of progesterone on the activity of GABA. It has been suggested that following a period of low allopregnanolone levels the administration of pregnanolone may cause an increase in anxiety and this anxiolytic response may be mediated by the amygdala. Healthy young adult females received a single injection of progesterone during the follicular phase of their cycles. The progesterone injection caused an increase in plasma progesterone and allopregnanolone levels to the levels normally observed during the luteal phase of the cycle and during early pregnancy, and resulted in increased levels of activity in the amygdala. The authors suggested that this increase in amygdala activity may be the mechanism by which progesterone modulates mood and anxiety (van Wingen et al. 2008).

A number of studies have used functional MRI to study brain function during particular tasks such as reading and mental arithmetic (Chapters 5 and 6). A final study worth noting is a study of the effects of a visual stimulus on brain activity as evaluated by MRI (Levin et al. 1998). Eight females and eight age-matched males were tested using blood-oxygen-level-dependent functional MRI. The subjects received alternations of 30-second periods of dark followed by 30 seconds of visual stimulation. The stimulus was delivered via goggles with light-emitting diodes, which flashed at a frequency of 8 Hz during the "stimulus on" periods. Responses were recorded from the left and right primary visual cortices. Analysis of the results showed that overall, the level of activation of the brains of the females was 38 percent less than the response of the males. In addition, the activation in the female subjects was symmetrical between the left and right areas, while the response in males was significantly greater on the right.

Conclusions

The following summarizes the main points to be drawn from Chapter 4.

1. Estrogen receptors are present in the brains of males and females. Activation of estrogen receptors may produce genomic responses, or nongenomic responses via membrane-bound receptors.
2. There is substantial evidence for interactions of estrogen with GABA, 5-HT, and dopamine systems. The nature of the interactions varies between brain regions and receptor populations.
3. Global measures of brain function have demonstrated female-male differences. EEG activity in females changes across the phases of the menstrual cycle and there is evidence for differences between females and males.

Bibliography and recommended readings

Abel, K. M., Y. O'Keane, R. A. Sherwood, and R. M. Murray. 1996. Plasma homovanillic acid profile at different phases of the ovulatory cycle in healthy women. *Biological Psychiatry* 39: 1039–143.

Azari, N. P., S. I. Rapoport, C. L. Grady, C. DeCarli, J. V. Haxby, M. B. Schapiro, and B. Horwitz. 1992. Gender differences in correlations of cerebral glucose metabolic rates in young normal adults. *Brain Research* 574: 198–208.

Benarroch, E. 2007. Endocannabinoids in basal ganglia circuits. *Neurology* 69: 306–09.

Best, S. E., P. M. Sarrel, R. T. Malison, M. Larnell, S. S. Zoghbi, R. M. Baldwin, J. P. Seibyl, R. B. Innis, and C. H. van Dyck. 2005. Striatal dopamine transporter availability with [^{123}I] β-CIT SPECT is unrelated to gender or menstrual cycle. *Psychopharmacology* 183: 181–89.

Biver, F., F. Lotstra, M. Monclus, D. Wikler, P. Damhaut, J. Mendlewicz, and S. Goldman. 1996. Sex difference in 5HT$_2$ receptor in the living human brain. *Neuroscience Letters* 204: 25–28.

Bossé, R., R. Rivest, and T. Di Paolo. 1997. Ovariectomy and estradiol treatment affect the dopamine transporter and its gene expression in the rat brain. I. *Molecular Brain Research* 46: 343–46.

Brenner, R. P., R. F. Ulrich, and C. F. Reynolds III. 1995. EEG spectral finding in healthy, elderly men and women — sex differences. *Electroencephalography and Clinical Neurophysiology* 94: 1–5.

Cosgrove, K. P., C. M. Mazure, and J. K. Staley. 2007. Evolving knowledge of sex differences in brain structure, function, and chemistry. *Biological Psychiatry* 62: 847–55.

Di Paolo, T., F. Bedard, and P. J. Bedard. 1989. Influence of gonadal steroids on human and monkey cerebrospinal fluid homovanillic acid concentrations. *Clinical Neuropharmacology* 12: 60–66.

Di Paolo, T., P. Falardeau, and M. Morissette. 1988. Striatal D-2 dopamine agonist binding sites fluctuate during the rat estrous cycle. *Life Sciences* 43: 665–72.

Espisto, G., J. D. Van Horn, D. R. Weinberger, and K. F. Berman. 1996. Gender differences in cerebral blood flow as a function of cognitive state with PET. *Journal of Nuclear Medicine* 37: 559–64.

Goldstein, M., and A. Y. Deutch. 1992. Dopaminergic mechanisms in the pathogenesis of schizophrenia. *FASEB* 6: 2413–21.

Grattan, D. R., and M. Selmanoff. 1997. Sex differences in the activity of γ-aminobutyric acidergic neurons in the rat hypothalamus. *Brain Research* 775: 244–49.

Guisti, M., L. Fazzuoli, D. Cavallero, and S. Valenti. 2002. Circulating nitric oxide changes throughout the menstrual cycle in healthy women and women affected by pathological hyperprolactinemia on dopamine agonist therapy. *Gynecological Endocrinology* 16: 407–12.

Herbison. A. E. 1997. Estrogen regulation of GABA transmission in rat preoptic area. *Brain Research Bulletin* 44: 321–26.

Johnson, M. H., and B. J. Everitt. 1995. *Essential Reproduction*. 4th Ed. Oxford: Blackwell Science.

Kaneda, Y., T. Ikuta, H. Nakayama, K. Kagawa, and N. Furuta. 1997. Visual evoked potential and electroencephalogram of healthy females during the menstrual cycle. *Journal of Medical Investigation* 44: 41–46.

Kaneda, Y., H. Nakayama, K. Kagawa, N. Furuta, and T. Ikuta. 1996. Sex differences in visual evoked potential and electroencephalogram of healthy adults. *Tokushima Journal of Experimental Medicine* 43: 143–57.

Koike, S., M. Sakai, and M. Muramatsu. 1987. Molecular cloning and characterization of rat estrogen receptor cDNA. *Nucleic Acids Research* 15: 2499–513.

Kritzer, M. F., and S. G. Kohama. 1998. Ovarian hormones influence the morphology, distribution and density of tyrosine hydroxylase immunoreactive axons in the dorsolateral prefrontal cortex of adult rhesus monkeys. *Journal of Comparative Neurology* 395: 1–17.

———. 1999. Ovarian hormones differentially influence immunoreactivity for dopamine β-hydroxylase, choline acetyltransferase, and serotonin the dorsolateral prefrontal cortex of adult rhesus monkeys. *Journal of Comparative Neurology* 409: 438–51.

Kuiper, G.G.J.M., E. Enmark, M. Pelto-Huikko, S. Nilsson, and J.-A. Gustafsson. 1996. Cloning of a novel estrogen receptor expressed in rat prostate and ovary. *Proceedings of the National Academy of Sciences, U.S.A.* 93: 5925–30.

Laere, K. V., K. Goffin, C. Casteels, P. Dupont, L. Mortelmans, J. de Hoon, and G. Bormans. 2007. Gender-dependent increases with healthy aging of the human cerebral cannabinoid-type 1 receptor binding using [18F]MK-9470PET. *NeuroImage* 39: 1533–41.

Laflamme, N., R. E. Nappi, G. Drolet, C. Labrie, C. Rivest, and S. Rivest. 1998. Expression and neuropeptidergic characterization of estrogen receptors (ERalpha and ERbeta) throughout the rat brain: Anatomical evidence of distinct roles of each subtype. *Journal of Neurobiology* 36: 357–78.

Levesque, D., and T. Di Paolo. 1988. Rapid conversion of high into low striatal D2-dopamine receptor agonist binding states after an acute physiological dose of 17 beta-estradiol. *Neuroscience Letters* 88: 113–18.

Levesque, D., S. Gagnon, and T. Di Paolo. 1989. Striatal D1 dopamine receptor density fluctuates during the rat estrous cycle. *Neuroscience Letters* 98: 345–50.

Levin, J. M., M. H. Ross, J. H. Mendelson, N. K. Mello, B. M. Cohen, and P. F. Renshaw. 1998. Sex differences in blood-oxygenation-level-dependent functional MRI with primary visual stimulation. *American Journal of Psychiatry* 155: 434–36.

Majewska, M. D. 1991. Neurosteroids: Endogenous bimodal modulators of the GABA$_A$ receptor. Mechanism of action and physiological significance. *Progress in Neurobiology* 38: 379–95.

Masonis, A.E.T., and M. P. McCarthy. 1995. Direct effects of the anabolic/androgenic steroids, stanozolol and 17a-methyltestosterone, on benzodiazepine binding to the γ-aminobutyric acid$_A$ receptor. *Neuroscience Letters* 189: 35–38.

Matsuura, M., K. Yamamoto, H. Fukuzawa, Y. Okubo, H. Uesugi, M. Moriiwa, M. T. Kojima, and Y. Shimazono. 1985. Age development and sex differences of various EEG elements in healthy children and adults — quantification by a computerized wave form recognition method. *Electroencephalography and Clinical Neurophysiology* 60: 394–406.

McEwen, B. S., E. R. de Kloet, and W. Rostene. 1986. Adrenal steroid receptors and actions in the nervous system. *Physiological Reviews* 66: 1121–88.

McEwen, B. S., J. M. Weiss, and L. S. Schwartz. 1968. Selective retention of corticosterone by limbic structures in rat brain. *Nature London* 220: 911–12.

Mermelstein, P. G., J. P. Becker, and D. J. Surmeier. 1996. Estradiol reduces calcium currents in rat neostriatal neurons via a membrane receptor. *Journal of Neuroscience* 16: 595–604.

Morissette, M., D. Biron, and T. Di Paolo. 1990a. Effect of estradiol and progesterone on rat striatal dopamine uptake sites. *Brain Research Bulletin* 25: 419–22.

Morissette, M., D. Levesque, A. Belanger, and T. Di Paolo. 1990b. A physiological dose of estradiol with progesterone affects striatum biogenic amines. *Canadian Journal of Physiology and Pharmacology* 68: 1520–26.

Niedermeyer, E., and F. Lopes da Silva. 1993. *Electroencephalography: Basic Principles, Clinical Applications, and Related Fields*. 3rd ed. Baltimore: Williams & Wilkins.

Nishizawa, S., C. Benkelfat, S. N. Young, M. Leyton, S. Mzengeza, C. de Montigny, P. Blier, and M. Diksic. 1997. Differences between males and females in rates of serotonin synthesis in human brain. *Proceedings of the National Academy of Sciences, U.S.A.* 94: 5308–13.

Pohjalainen, T., J. O. Rinne, K. Nagren, E. Syvalahti, and J. Hietala. 1998. Sex differences in the striatal dopamine D$_2$ receptor binding characteristics in vivo. *American Journal of Psychiatry* 155: 768–73.

Ramirez, V. D., J. Zheng, and K. M. Siddique. 1996. Membrane receptors for estrogen, progesterone and testosterone in the rat brain: Fantasy or reality. *Cellular & Molecular Neurobiology* 16: 175–98.

Rissman, E. F., S. R. Wersinger, H. N. Fugger, and T. D. Foster. 1999. Sex with knockout models: Behavioral studies of estrogen receptor alpha. *Brain Research* 835: 80–90.

Rubinow, D.R., P. J. Schmidt, and C. A. Roca. 1998. Estrogen-serotonin interactions: Implications for affective regulation. *Biological Psychiatry* 44: 839–50.

Sar, M., W. E. Stumpf, and M. L. Tappaz. 1983. Localization of ^{3}H-estradiol in preoptic GABA neurons. *Federal Proceedings* 42: 495.

Simerly, R.B., C. Chang, M. Muramatsu, and L. W. Swanson. 1990. Distribution of androgen and estrogen receptor mRNA-containing cells in the rat brain: An in situ hybridization study. *Journal of Comparative Neurology* 294: 76–95.

Small, J. G. 1967. Central Association of Electroencephalographers. *Electroencephalography and Clinical Neurophysiology* 23: 589–94.

Smith, Y. R., J-K. Zubieta, M. G. del Carmen, R. F. Dannals, H. T. Ravert, H. A. Zacur, and J. J. Frost. 1998. Brain opioid receptor measurements by positron emission tomography in normal cycling women: Relationship to luteinizing hormone pulsatility and gonadal steroid hormones. *Journal of Clinical Endocrinology and Metabolism* 83: 4498–505.

Toran-Allerand, C. D., M. Singh, and G. Setalo Jr. 1999. Novel mechanisms of estrogen action in the brain: New players in an old story. *Frontiers in Neuroendocrinology* 20: 97–121.

Tseng, A. H., J. W. Harding, and R. M. Craft. 2004. Pharmacokinetic factors in sex differences in Delta 9-tetrahydrocannabinol-induced behavioral effects in rats. *Behavioural Brain Research* 154: 77–83.

Van Laere, K., K. Goffin, C. Casteels, P. Dupont, L. Mortelmans, J. de Hoon, and G. Bormans. 2007. Gender-dependent increases with healthy aging of the human cerebral cannabinoid-type 1 receptor binding using [^{18}F]MK-9470 PET. *NeuroImage* 39: 1533–41.

Van Wingen, G. A., F. van Broekhoven, R. J. Verkes, K. M. Petersson, T. Backstron, J. K. Buitelaar, and G. Fernandez. 2008. Progesterone selectively increases amygdala reactivity in women. *Molecular Psychiatry* 13: 325–33.

Volkow, N. D., G-J. Wang, J. S. Fowler, R. Hitzemann, N. Pappas, K. Pascani, and C. Wong. 1997. Gender differences in cerebellar metabolism: Test-retest reproducibility. *American Journal of Psychiatry* 154: 119–21.

Weiland, N. G. 1992. Glutamic acid decarboxylase messenger ribonucleic acid is regulated by estradiol and progesterone in the hippocampus. *Endocrinology* 131: 2697–702.

Zheng, J., and V. D. Ramirez. 1999. Purification and identification of an estrogen binding protein from rat brain: Oligomycin sensitivity-conferring protein (OSCP), a subunit of mitochondrial FOF1-ATP synthase/ATPase. *Journal of Steroid Biochemistry & Molecular Biology* 68: 65–75.

chapter 5

Perception and cognition

In the early days of experimental psychology, B. F. Skinner gave us the "black box" of behaviorism (Skinner 1969). In the world of Skinner, the stimulus and the response were everything. The black box, the organism doing the responding, was simply a processing mechanism. The sensory systems were important for perceiving the stimulus; motor systems were necessary for generating a response. Cognitive elements were simply not an issue of interest. The black box was not required to think.

It did not take long for the cracks in this radical form of behaviorism to begin to appear. In animal learning laboratories, researchers observed that rats, when unable to make the conditioned response, would come up with novel ways of obtaining their reward. In one particularly amusing example, the tension on the lever in the Skinner box was increased until it was difficult for the rat to depress it far enough to receive the reward. Rats, previously trained to bar-press for food reward, were placed in the Skinner box. The animals' attempts at responding were recorded as they tried in vain to earn their reward. One particular rat solved the problem. After a period of furious, frustrated attempts at bar pressing, the rat turned around, backed onto the apparatus and using both hind legs depressed the lever far enough to receive his reward (R. Champion, personal communication). It is difficult to argue that some form of cognitive processing had not occurred in the rat's brain.

The processes involved in sensation, perception, and cognition are the domain of experimental psychology and psychophysics. The experimental psychologist faces some of the greatest challenges in the world of scientific experimentation. Good experimental design requires rigorous control of the experimental conditions. For example, a "simple" chemistry experiment to establish the boiling point of a liquid, such as a sugar solution, requires that a number of observations be made under the same conditions. This is an apparently straightforward requirement, at least until you think about how to go about it. First, consider the materials and methods. You will need a heat source, a container for the liquid, and a method of measuring the temperature of the liquid. These must be the same for every measurement that is made. The conditions in the laboratory must also be identical. Ambient temperature and humidity must be controlled. Atmospheric pressure must also be considered; not only do the experiments have to be conducted at the same altitude but also under identical weather conditions.

There is the experimental "subject," the liquid to be measured. If the liquid is mixed fresh before each measurement, all of the mixing procedures must be carefully controlled and the constituent parts must come from the same source. If the liquid is to be stored between measurements, then variability can be introduced through aging or contamination. Either way, keeping the experimental subject constant can be difficult.

Now, imagine the complexity of the experiment multiplied a billion-fold. Instead of a simple sugar solution, make the subject a 34-year-old human female. She has a 34-year plus 9-month (approximate gestation period) history of development and maturation that is unique to her. She also has her own genetic make-up (unless she is a monozygotic twin) and 34 years of experience of the world. She really is a "one-of-a-kind" subject. But, if she takes part in an experiment, she will become one of the x number of subjects whose results will be pooled to yield descriptive statistics such as means and standard deviations. The kind of experimental rigor required for psychophysical experiments is staggering, and has become a trademark of experimental psychology.

The material covered in this chapter comes from a range of psychophysical experiments including studies of sensory systems, cognitive processes, learning, and memory. Some of the experimental results that we will cover come from simple "pencil and paper" tasks; other results come from experimental paradigms using state-of-the-art computing systems. In either case, the experiment will have started with the delivery of a constant set of instructions to the experimental subject. To people unfamiliar with experimental psychology, if often comes as a surprise that the delivery of instructions to the subjects, either before or during the experiment, is always made in a precise and replicable manner. In the mid-1900s, when experimental psychology was a young science, the instructions to the subjects were carefully written by the experimenters and read to the subjects from a prepared script. The language usage was checked and rechecked to avoid ambiguities or leading instructions that could suggest to the experimental subject what was "expected" of them. Consider the following spoken instruction, "When the buzzer sounds, please turn the paper over and circle every word not printed in red ink." What has the experimental subject just been asked to do? Has she been asked to circle every single word on the page printed in an ink color other than red? Or has she been instructed to circle every instance of the word "not" that is printed in red ink? Many researchers still read scripts to experimental subjects, but now recordings and short videos of the instructions are more frequently used.

By giving consistent instructions to experimental subjects, the experimenter hopes to exclude a potential source of variation in the results. First-year university students are a group commonly used as participants in psychophysical experiments. Using first-year students is, in itself, a means of controlling some of the possible variables. There will be some consistency in intelligence and years of education. The age range of the participants will also be fairly restricted, which may or may not be an advantage. Unfortunately, one area of variation in performance that has been consistently overlooked is sex differences. This omission seems particularly strange in a discipline noted for its experimental control.

IQ: When is a difference not a difference?

Before starting to examine the data from experiments on cognition and perception, it is important to revisit the question of intelligence. This has to be one of the most emotionally and politically charged questions in sex and gender issues. It is almost always reduced to the question of "higher or lower," and the proponents of this kind of question usually want an absolute number. This is such an odd approach given the vast literature demonstrating that intelligence is based on a wide range of factors, as reflected in the diversity of tasks that are assessed by an IQ test. It is not surprising that individual people excel on different kinds of tasks, nor is it surprising that females as a group are better at some tasks and males are better at others. The final IQ test score is made up of a number of scores from the different subcategories of the test battery, and it will reflect some aspects of the individual's cognitive abilities. IQ tests certainly have their place, particularly when trying to understand an individual with apparent cognitive abnormalities or one who displays signs of cognitive decline. Where they are not as useful is in trying to pinpoint a normal, healthy individual (or group of individuals) to an exact location on a numerical scale. A good example of the shortcomings of this kind of assessment is found in the data on university entrance test results. Males usually score higher on the university entrance tests, but females get higher grades.

Evidence is presented in Chapter 3 to support the idea that the larger the brain, the higher the IQ. This evidence applies to males and females and is interesting, as far as it goes. Clearly, however, volume alone is not enough to explain individual differences in intelligence. Until relatively recently, the only tools to examine intelligence were the standardized IQ test batteries. There are a number of reliable, well-validated IQ tests that are available in a number of countries and languages. These tests provide a numerical guide to the cognitive ability of the individual. They do not, however, provide any insight into the workings of the brain whose IQ is being tested.

The question of the how brains function to provide "intelligence" has barely been touched upon. The neurotransmitter acetylcholine has been called the transmitter of cognition, largely because it is the first neurotransmitter to become deficient in Alzheimer's disease. One school of thought has held that specific (as yet unspecified) regions of the brain are the source of intelligence. Another school has held that it is the connections between the areas of the brain, the "neural circuitry," that forms the basis of cognition and intelligence. A third approach, not surprisingly, suggests that it is a combination of regional specificity combined with neural circuitry that constitutes cognitive function. While the different approaches are interesting, they still do not approach the question of what the brain is doing that produces intelligence. Several recent approaches to this question have examined the way specific brain processes are correlated with intelligence, as measured by the standard IQ tests. One particular group has been evaluating energy expenditure in the brain in respect to particular types of tasks. From this work they have proposed "the neural efficacy hypothesis of intelligence" (Haier 1992). According to this hypothesis, the greater the intelligence of an individual, the less hard the brain has to work. Imagine the brain as a swimmer. The highly skilled athlete cruises up and down the pool effortlessly while the poor swimmer puffs and pants and struggles to complete even one lap. Working less hard does not shed light on what cognition is, but it does suggest that whatever it is, if you have extra energy left over, you can do more of it. This suggestion leaves large explanatory gaps to be filled, but at least it provides a starting place where researchers can look for supporting evidence. If this hypothesis is correct, then not only would you expect people with higher intelligence to expend less energy on cognitive tasks, you would also expect that individuals would expend less energy processing tasks for which they have greater ability.

It has recently been reported that females and males demonstrate "neural efficacy" for the tasks in which they are known to excel, that is, verbal and visuospatial tasks, respectively (Neubauer 2005). In this study, females (n = 35) and males (n = 31) were screened for general intelligence using an established IQ test. Following screening, each subject completed four experimental tasks while their EEG was recorded. These tasks were "same or different" judgments: (1) the emotion of two simultaneously presented facial expressions; (2) nonemotional control judgment, the sex of two simultaneously presented faces; (3) verbal task, the meaning of two simultaneously presented words; and (4) visuospatial, the type of rotation of two simultaneously presented arrows. Each testing session lasted approximately two hours with short (five-min.) breaks. When the results of the tests were analyzed, the "IQ-activation" relationship differed depending upon the task being performed. Females showed less activation during the performance of the verbal matching task, while males showed less activation during the visuospatial task. The researchers also noted that the sex differences in neural efficacy were most obvious in the cortical areas related to verbal matching for females, and the visuospatial task for males.

In another study, the biochemical basis of intelligence was examined using magnetic resonance spectroscopy (MRS), an imaging technique that allows the activity of brain chemicals to be measured as networks in the white matter (the nerve fibers connecting the neurons) are activated. A particularly interesting neurochemical for this kind of study is N-acetylaspartate (NAA), a substance produced in the mitochondria of neurons and used to measure the integrity of those neurons. For this study, 10 females and 17 males, mean age 25 years, were first screened for IQ using a standard IQ test battery. There was no significant difference in mean IQ between the two groups. Each subject was then scanned on another occasion within two weeks of the day of cognitive testing. Three brain regions were selected for specific analysis: the left frontal area, the right frontal area, and the left occipito-partietal area. As would be expected, the total brain volumes were significantly greater in males than in females. Looking at the specific regions of interest, a combination of higher levels of NAA in the occipito-parietal area and lower NAA in the left-frontal area was the best predictor of IQ in females but not in males. Further analysis by IQ test subtests revealed significant correlations between the picture completion task, symbol search, and arithmetic and left occipito-parietal area NAA. Block design was negatively correlated with NAA in the left frontal area in females. The correlation between the two different brain regions is taken as evidence for the importance of neural circuitry in female intelligence. There were no significant regional correlations with IQ in males. In their discussion the authors suggested that the neuronal efficiency hypothesis, in light of the regional NAA correlations in females but not males, may explain why female and male mean IQ is similar despite the 10 percent smaller size of the female brain.

Cognitive performance during pregnancy is an area where anecdotal evidence abounds. Our grandmothers have told us that we can expect to be emotional, forgetful, and illogical while pregnant, or at least that that is what our male partners will expect. The evidence for cognitive changes, however, has not been impressive. A study of 198 pregnant females and 132 nonpregnant controls yielded mixed results (Crawley 2002). One problem always confronting researchers in studies of pregnancy is finding an appropriate nonpregnant control group. For this study, the experimenters decided to use female university students who were living away from home for the first time, the rationale being that both groups were undergoing a period of major transition in their lives. The pregnant subjects were recruited from an antenatal clinic, the nonpregnant females from a seminar they attended within four weeks of moving out of home. The mean ages of the two groups, pregnant = 29 years, and nonpregnant = 19 years, were significantly different. For this study, the participants were first asked to report any recent changes in cognitive experience. They were then asked to complete an inventory of changes in four areas: general memory, concentration, clarity of thought, and prospective memory. In response to an open-ended question about recent cognitive experiences, the responses of the two groups were very similar: only three females in each group reported cognitive changes; about 25 percent of both groups reported emotional changes. In response to the cognitive function checklists, the majority of both groups reported no change. For the subjects who did report a change, the groups responded in opposite directions, for the pregnant group the change was in the "worse" direction, while the change was in the "better" direction for the control group. Following this experiment, the experimenters conducted a second experiment, using a different protocol and more closely matched groups. Thirteen females who already had one child and were in the second trimester of pregnancy, mean age 33 years, were in the experimental group. Thirteen females in the nonpregnant group, mean age 34 years, were recruited from a postnatal group more than one year following their first pregnancy. The mean ages of the children of each group was similar, around two years old.

The subjects completed the Cognitive Failures Questionnaire (CFQ) and reported cognitive "slips" in the past four weeks. Testing occurred on three separate occasions: second trimester (20–24 weeks); third trimester (33–36 weeks); and five to six months postpartum. The females' partners completed the CFQ-for-others at the same testing times. The CFQ scores for the pregnant group were significantly higher (i.e., more impaired) than the scores for the nonpregnant group; however, the scores of the partners for the two groups did not differ. The interesting point to take from this study is that pregnant women do not see cognitive changes as a particularly salient part of their experience, only reporting perceived changes when directly questioned, while their partners do not perceive any changes at all.

Visual perception

One of the earliest visual perception tasks to reveal female-male differences in spatial ability was the "rod and frame" task described by Witkin et al. in 1954 (referenced in Silverman et al. 1973). This is a simple task, requiring only the adjustment of a rod within a frame to earth vertical (Figure 5.1). Intuitively, it should be easy. The subject sits in a darkened room with only an illuminated frame and rod visible. The rod and frame may be moved independently so that they can be tilted in the same direction away from vertical or may be tilted in opposite directions from vertical. Usually, the trials are a randomized mix of same and different directions of tilt. Silverman et al. (1973) tested 15 female and 15 male university students, aged between 18 and 22. Their subjects were seated in a darkened room and instructed to set the rod to be parallel with the walls of the room. The subjects' heads were fixed in a vertical position using a

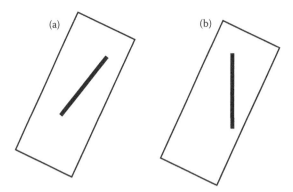

Figure 5.1 Schematic representation of the Rod and Frame Test. The subject is instructed to position the rod vertically within the tilted frame: (a) unsolved task; (b) correct solution.

padded headrest. The results of eight trials were averaged to give a mean error score. The error scores for the females were significantly higher than the error scores for the males.

Another area of visual perception where females have traditionally performed less well than males is in tasks involving mental rotation of images. These tasks usually employ a target image, often composed of blocks (Figure 5.2). The subject is required to match the target image to a test image, a rotation of the target image, which is presented with other test images such as mirror images of the target, which have been rotated in three-dimensional space. In order to complete the task, experimental subjects apparently perform a mental rotation of the target, looking for a match to a test image. In 1995, Desrocher et al. measured event-related potentials (ERPs) associated with mental rotation performance. The subjects were university students, 10 females and 10 males, with mean ages of 21 years and 22 years, respectively. All of the subjects were right-handed. Two kinds of images were used in the study, letters and simple abstract designs, presented on a computer screen. The subjects were instructed to look at a target image on the left side of the computer screen and to judge whether or not the image on the right side of the screen was a rotation of the target image. They signaled their decision by pressing a button with their index finger (the hand for "same" and "different" varied across subjects). In order to ensure the accuracy of the subjects' responses, they were given 80 practice trials before the experimental trials (160) began. Although there were no reaction time (RT) differences between females and males, several other differences in processing emerged. The female subjects began their analysis earlier, as indicated by a shorter latency for

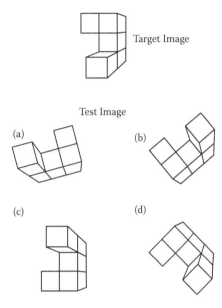

Figure 5.2 Mental rotation. The subject is required to choose the two test figures that are rotations of the same figure. Correct answers: (a) and (d).

the N400 wave. In addition, females showed a greater positivity than males in the wave-forms between approximately 400 and 800 msec, which are associated with the later stages of stimulus evaluation and the early stages of mental rotation. Finally, the timing of the processing in the female subjects differed for the abstract designs and the letters. When the target was an abstract design, processing began earlier if the amount of rotation was large (i.e., a large angle) than if the test figure had only been rotated by a small amount. For letters, however, large angle processing began later than small angle processing. This difference between real and abstract images seems to be an important issue and we will return to it later.

Silverman et al. (1993) conducted a series of four experiments examining the relationship of performance on a similar mental rotation task to phases of the menstrual cycle. The subjects were recruited mostly from a student population. The mean ages of the different groups was not specified, however. In the first of the four experiments the subjects were 158 females and 105 males. The menstrual cycle testing was conducted as a between-subjects design. The female subjects were divided into three groups: (1) not on the contraceptive pill (n = 101); (2) currently taking the pill (n = 42); and (3) taking the pill but in the "off" (menstrual) phase of the pill cycle (n = 15). The performance of the male subjects was significantly better than the performance of the female subjects. Female pill users also performed significantly better than non-pill users. When this result was broken down further, it was found that the performance of pill users in the "off" phase was equivalent to the performance of the male subjects. In the second experiment, a within subjects design, female contraceptive pill users (n = 27) were tested during their "on" and "off" phases. Female non-pill users (n = 23) were tested during their menstrual phase and during the luteal phase (days 16–23 for the majority of subjects). The performance of the menstrual phase/"off" pill subjects was significantly better than the performance of the luteal phase/"on" pill subjects. The aim of experiment 3 was to determine if the results obtained in experiments 1 and 2 were specific to the mental rotation task or could also be found using other tests of cognitive function. In this experiment the results on the mental rotation task were compared with the results on two "control" tasks, Digit Symbol, which measures a number of variables including visual-motor coordination, and anagram solving. The effects found in experiments 1 and 2 were replicated for the mental rotation task only. Finally, experiment 4 used a variation of mental rotation, the space relations task, to examine further the mental rotation effects found in the first three experiments. The space relations task required the subject to look at a two-dimensional target pattern and decide which of a number of target three-dimensional objects could be constructed by folding the test pattern. One hundred and sixty two subjects took part in the experiment; males (n = 42), females (menstrual/"off" phase, n = 26) and females (luteal/"on" phase, n = 94). In both tasks the males performed significantly better than the females. For the mental rotation, menstrual/"off" phase females performed significantly better than luteal/"on" phase females (Figure 5.3). The responses of the two groups of females were equal for the space relations task. Another study of mental rotation and the menstrual cycle yielded similar results (Moody 1997). Males performed better than females overall. The female subjects scored significantly better during the menstrual phase than during the luteal phase, and the performance of females in the menstrual phase was not significantly different from the performance of the males.

Although the biological basis of the male advantage in the mental rotation task is not understood, it has been suggested that females may use different processing strategies than males, and this may, at least in part, contribute to the difference in performance. An fMRI study recorded the brain activity of females (n = 13) and males (n = 12) during a mental rotation task (Butler et al. 2006). In their introduction, the authors noted that previous fMRI studies of mental rotation have given variable and conflicting results, possibly based on the inconsistency of the mental rotation tasks used. For their experiment, Butler and colleagues used a computerized mental rotation task that was optimized and validated for use during fMRI. The task incorporated different levels of difficulty, that is, different amounts of image rotation ranging from 40° to 160°. Accuracy and response time were recorded on all trials. There were two test conditions,

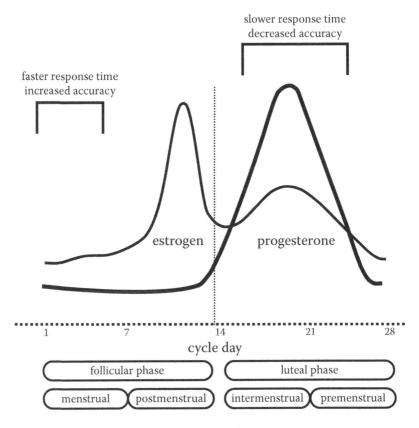

Figure 5.3 Mental rotation and the menstrual cycle.

rotation (same or different figures, between 40° and 160°) and compare (same or different figures, no rotation). For both females and males, accuracy and reaction time were correlated with the degree of rotation of the figure. In both females and males, the prefrontal and parietal cortices were activated bilaterally, as were the temporal-occipital regions, visual association cortices, and diencephalons. In females, the greatest correlation between mental rotation and brain activity was in the left middle frontal gyrus. In males, the only significant correlation between activity and performance was in the left posterior insula at the border of the claustrum in an area that receives input from the vestibular system. Interestingly, the magnitude of the effect was proportional to the magnitude of rotation of the image, suggesting to the experimenters that males may have been using a more literal rotation strategy than females — for example, imagining they were tilting until they were aligned with the image. Overall, the results of this study led the authors to conclude that females and males used quite different strategies in their approach to determining the nature of the rotated images. The greater activity changes in the higher areas of the association cortices, they suggest, are consistent with females using a top-down processing strategy. Males, on the other hand, by using the visual-vestibular network may be using a more basic, unconscious bottom-up strategy.

A number of variables in the experimental design of mental rotation tasks, including the nature of the target objects and the instructions to the subjects, have been studied. It has been suggested that the nature of the test objects (real or abstract) may be important. One study used pictures of mannequins, dressed either as females or males, instead of abstract shapes, as the test figures (Richardson 1994). Each mannequin had a black disk in one hand and a white disk in the other hand. The mannequins were presented at different orientations, and the task of the subject was to determine whether the black disk was in the mannequin's left or right hand. When the male mannequins were presented to one group of females and one group of males, the task was labeled as a "test of spatial ability." When the female mannequins were presented to two different groups of females and males, the task was labeled as a "test of personal empathy." On both tasks the males performed better than the females. In contrast, the instructions given to participants in another mental rotation task significantly changed the performance of the subjects (Sharps et al. 1994). Using the standard target and test figure paradigm, half of the 36 females and 36 males were given instructions that emphasized the spatial nature of the task while the other participants were given instructions that did not mention spatial ability. When the results were analyzed, the subjects in the "spatial task" condition showed the usual males-better-than-females response pattern. The performance of the female and male subjects who received the "nonspatial" instructions was the same, however.

It has been suggested that the difficulty of three-dimensional rotations might create a performance bias in favor of males. The results of a study comparing two-dimensional (hard and easy) and three-dimensional figures revealed that males performed better on all three tasks, although there was less male advantage on the "easy" two-dimensional task (Collins and Kimura 1997). Interestingly, when the task employed real, three-dimensional models of the printed three-dimensional figures used in most mental rotation tasks, the sex difference disappeared (McWilliams et al. 1997).

In another type of spatial task, this time remembering locations in space, females showed a definite advantage (McBurney et al. 1997). The performance of 46 females and 57 males, mean age 20 years for both groups, was compared on a standard mental rotation task and a spatial memory task. The spatial memory task used a commercially available board game that required players to remember the location of previously viewed images. As expected, the males performed significantly better on the mental rotation task; however, the females were superior to the males in remembering spatial locations. The authors suggested that superior location memory may be related to "evolutionary adoptedness." In a hunter-gatherer society, it was essential for the female gatherers to remember the location of food.

Another experimental result, which could conceivably relate to different strategies for the hunters and the gatherers, has been reported for geographic knowledge and directions (Dabbs et al. 1998). This experiment had 104 females and 90 males complete a series of tasks measuring geographical knowledge and the ability to give directions from a map. The results of the direction-giving part of the experiment were particularly interesting. While the males gave directions in Euclidian space including compass points and distances, the females gave more personal directions including left and right turns and landmarks. With a slight stretch of the imagination, one can envisage how Euclidian directions would advantage the long-distance hunter, while the "stay-near-camp" gatherer would benefit more from references to turning directions and landmarks.

From the material presented above, it is clear that females do not perform as well as males on some tests of spatial ability. Does this mean that there are some areas of performance — for example, jobs requiring high spatial ability — that are unsuitable for females? This is, not surprisingly, a question of great interest to institutions such as the Air Force. Interestingly, this question was addressed, and answered, in 1982 (McCloy and Koonce 1982). A group of 103 U.S. Air Force Academy Cadets (52 female, 51 male) were included in the study. The cadets were given a battery of tests that included spatial orientation as well as other aspects of cognitive and motor functioning. The males performed significantly better than the females on the mental rotation task as well as the psychomotor and flight simulator tasks. A year later, the cadets who were still at the Air Force Academy were retested. This time, the test procedure required the cadets to reach a specified performance criterion on each task. The females required more trials to reach criterion but the same standard of performance was achieved. The authors suggested that in designing tests to predict performance of certain skills, such as flying a high-performance aircraft, the sex of the test subjects will be an important variable to be included in the test design.

The Stroop task is a visual task that requires both reading and color recognition. The task, until one has tried it, appears quite straightforward. The names of colors, for example, red and green, are displayed to subjects in letters colored either to match the named color ("green" in green letters) or in a color different from the named color ("green" in red letters). The task of the subject is to press a button corresponding to the named color or to press a button corresponding to the color of the letters and ignore the color name. Instead of a button press, a verbal response may be used. The variables measured are accuracy and reaction time (RT). The "Stroop effect," the interference caused by nonmatching words and letter colors, is apparent as increased RT and decreased response accuracy. The literature on female-male differences in the Stroop task performance is mixed. Some authors find a difference; some do not. The following two studies are representative of the kinds of contradictory results found in the literature. A study of 8 female and 8 male undergraduates reported that although the response (a button press) accuracy was equal for females and males, RT for males was longer (Mekarski et al. 1996). A larger study of 69 females and 59 males, also undergraduates, reported that when the task, which required a verbal response, was completed in a "relaxed" atmosphere, RT for females was longer than for males, but the responses of the females were more accurate (von Kluge 1992). When the conditions of the task were changed and the importance of performance was stressed, however, the results changed. The RT of the females decreased to match male RT, but the greater accuracy of the female subjects was maintained.

Weekes and Zaidel (1996) designed a series of studies to try to determine which factors are most important in testing the Stroop effect. The factors in the experiments included: (1) mode of visual presentation, bilateral or unilateral; (2) mode of response, verbal or manual; (3) separation of color and word; and (4) phase of menstrual cycle, menstrual or mid-luteal. The subjects were 146 right-handed, native English speakers. Equal numbers of females and males were used for each experiment. The females were aged between 20 and 35 years; the age range of the male subjects was not stated. The authors reported that factors 2 and 3 were not crucial to the experimental outcome, although the Stroop effect was somewhat weaker for manual rather than verbal responses. There was a clear hemispheric advantage, or disadvantage, with the left hemisphere showing a greater response decrement (Stroop interference) than the right hemisphere. In addition, there was an interaction with the day of the menstrual cycle. When the left hemisphere effects were further analyzed, the difference was significant only for males and for females in the menstrual phase when estrogen was low (Weekes and Zaidel 1996). Menstrual cycle effects have also been reported by Lord and Taylor (1991), who tested 50 female subjects every seven days for four weeks. The authors reported that the accuracy was low during the menstrual phase, peaked at midcycle, then dropped again as the next menstrual phase approached (Figure 5.4).

Object recognition tasks use a more "natural" form of stimulus that may or may not have some kind of emotional value for the viewer. The objects presented can range from images of faces and babies to garden implements and pieces of furniture. The variables measured will be either RT and/or accuracy. In a chair-recognition task, for example, following the tachistoscopic presentation of chair images, the subjects were required to

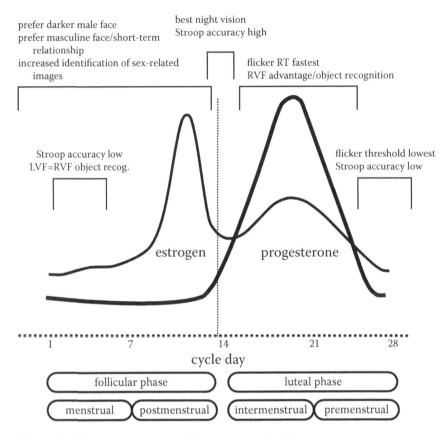

Figure 5.4 Visual perception and the menstrual cycle.

select the previously presented chairs from an array of chairs (Bibawi et al. 1995). The accuracy was the same for females and males. For males, the accuracy of their choices was also the same for left visual field (LVF) and right visual field (RVF) presentations. For females, however, LVF and RVF were equal only during the menstrual phase. During the mid-luteal phase there was a significant RVF advantage. In another study, photographs of urban scenes familiar to the subjects were used (Hamel and Ryan-Jones 1997). All of the scenes were viewed on a computer screen. For some of the presentations, the scenes were high-resolution photographs. On other presentations, the scenes were distorted to reduce the amount of detail. The male subjects identified the photographic and the distorted scenes with equal accuracy. For the female subjects, their accuracy for the photographic images was significantly greater than their accuracy for identification of the degraded images.

The human face raises some particularly interesting issues in object recognition. Overall, the faces of females and males differ in appearance. The facial skin of females tends to be lighter than for males. The skin also has a different texture. In the mature male, partial regrowth in shaved areas, the "5 o'clock shadow," also provides additional coloration to the face. The overall shape of the face also differs between females and males, with smaller-featured, symmetrical shapes being seen as more feminine. The responses of females and males to viewed faces also differ. For a study on skin tone preference (Feinman and Gill 1978), the subjects, 482 female and 549 male, white undergraduates, were asked to indicate their preferred skin color for the opposite sex. The female subjects reported a significantly greater preference for darker skin tones, while for males the preference was for lighter skin tones. In another study the female subjects were asked to make a preferred face choice (skin tone was the variable) at different phases of the menstrual cycle: subjects in the follicular phase showed a significant preference for darker male faces. Subjects in the luteal phase did not show a preference for face color, nor did women taking oral contraceptives (Frost 1994).

In 1999, it was suggested that the preference for facial characteristics changes with the probability of conception (Penton-Voak et al. 1999). In a study of 39 women with a mean age of 21 years, the subjects with the greatest probability of conception (days 5–14) preferred the most masculine faces in a stimulus array. In a second experiment, 65 subjects were allowed to use a computer program to manipulate images of male faces, making them more or less masculine. The subjects were asked to indicate which face they preferred for a long-term relationship and which face they preferred for a short-term sexual relationship. There was no difference between the phases of the menstrual cycle for the "long-term relationship" preference. For the short-term sexual relationship, however, the more masculine face was preferred when the probability of conception was high.

The meaning of the stimulus has also been shown to affect response patterns across the menstrual cycle. A study of 16 female subjects used tachistoscopic presentations of images with different meanings: sex (nude males), maternal (babies), body care (for example, a hairbrush), and neutral (nonsense syllables) (Krug et al. 1994). The subjects were tested at three stages of their menstrual cycles: menstrual, pre-ovulatory and midluteal. The ability to recognize nonsense syllables did not change across the menstrual cycle; however, the ability to recognize meaningful stimuli did. In the pre-ovulatory stage, there was an increase in the ability to identify stimuli in the sex category. There was also an increase in the number of incorrect identifications of sex stimuli. In the mid-luteal phase, there was an increased sensitivity to babies. Sixteen women taking oral contraceptives were tested in the same study. Their responses to pictures did not change across the cycle; however, their overall performance was less accurate during the menstrual phase. In an evoked potential study using similar stimuli, it was reported that only the P300 wave changed across the menstrual cycle, increasing in response to babies and males in the mid-luteal phase (Johnston and Wang 1991). The response to babies and males was highest in the mid-luteal phase. When the subjects were asked to rate the pleasantness of the slides they had just seen, the highest ratings for pleasantness were given during the pre-ovulatory phase.

Other changes in visual function across the menstrual cycle have also been reported. Generally, visual acuity has been reported to vary across the cycle; night vision has been reported to be best at midcycle, response to light flicker has the shortest latency in the luteal phase, and the threshold frequency to detect flicker has been reported to be lowest premenstrually (Figure 5.4; see Walker 1997 for a review).

Color: Perception and preference

Color preferences are surprisingly constant across the Western population although the "in" colors in fashion change with the seasons. When asked to name their favorite color, the majority of females and males will say "blue." When asked to choose a favorite color from a set of color swatches, the majority of people will also make a selection from the blue portion of the spectrum. The second choice for females is purple, and when choosing from swatches, many females select a blue that is on the purple side of the selections. Experimental evidence for sex differences in color preferences has been inconclusive. In a recent study, 208 subjects aged 20 to 26, including 110 females and 98 males, were tested for color preference using a forced-choice selection task. Sets of two small colored rectangles were presented sequentially in the center of an otherwise neutral computer screen. The subjects were instructed to use the mouse to select their preferred color as quickly as possible. Subjects participated in three separate experiments, each with the same each colors, using different values of saturation and lightness. Preferences were consistent for hue regardless of saturation and lightness of the presented color swatches. For females, the preference curves rose sharply in the in the reddish-purple range and remained high, then dropped steeply in the yellow-green range. The preference curve rose less steeply for males, with a wider preferred area in the blue-green range. There was more variability in the females' preferences and they were also more stable over the three experiments (Hurlbert, Ling 2007). When the authors examined reaction times they found that male reaction times were significantly faster than female reaction times and that for both females and males reaction times were significantly faster for "bluish" colors compared with "yellowish" colors. The faster reaction time is interesting in terms of the physiology of color vision, suggesting that the sensitivity to blue is greatest for both sexes. The neuronal mechanisms encoding color are composed of sets of cones (the color receptor cells in the eye), which are specific for either blue-yellow or red-green color coding. In making color preference choices, both females and males selected blue (blue-yellow cone opponent contrast) as the preferred color. But interestingly, both groups biased this preference toward one of the colors from the red-green cone opponent contrast. In fact, they chose opposite ends of the red-green contrast. Females biased their preferences toward the red end, and males biased their preferences toward the opposite green end. The female bias is consistent with the long-held belief that little girls prefer pink.

Colors are often described as either "warm" (red-orange-yellow) or "cool" (purple-blue-green). Preferences for warm and cool colors have been demonstrated to vary across the menstrual cycle, with warmer colors preferred in the luteal phase compared with the follicular phase (Kim and Tokura 1997). Color discrimination has also been reported to vary with the menstrual cycle, with color discrimination being most accurate at ovulation (Giuffre et al. 2007).

Auditory perception

The visual system is not the only sensory system where sex differences have been found. The visual system is only one of six sensory systems, and although vision receives the most attention, the other sensory systems are also vital for cognitive function. The auditory system is the second most-studied sensory system. Studies of auditory psychophysics range in complexity from measuring the ability to detect a single tone to the analysis of complex musical pieces.

A study of 24 female and 24 male undergraduates examined the subjects' ability to distinguish tones and series of tones presented binaurally or as a dichotic listening task (McRoberts and Sanders 1992). Half of each group were music majors or people with formal musical training (experienced); the other half of each group were people with no formal musical training (naïve). The subjects were given three tasks. The first task, the Seashore Pitch Discrimination Test, required the subject to listen to 50 pairs of tones and after each pair to report whether the second tone was of higher or lower pitch than the first tone. The second test, the Seashore Tonal Memory Test, required subjects to listen to 30 pairs of tonal sequences ranging in length from three to five tones per sequence. In each pair of sequences one note was different. After listening to each pair, the subject had to report which tone (first, second, third, fourth, or fifth) was different. The third test was the Fundamental-Frequency Contour Test. For the test the subjects listened to a continuous tone with changing pitch (contour). The pitch of the tone could rise, fall, rise then fall, or fall then rise. The subjects had to decide which contour occurred on each trial. The presentation of tasks was either binaural or dichotic, with the subject instructed to attend to the tone presented to either the right or left ear. Analysis of the results showed that the musically experienced subjects performed better than the musically naive subjects. For the binaural tasks, the males were more accurate than the females. Overall, there was a left-ear advantage for the dichotic tasks, and females and males performed equally well.

Still on the musical theme, there is evidence to suggest there are sex differences in the perception of loudness, which vary depending upon whether you ask the listeners to rate "loudness" or "annoyance" (Fucci et al. 1994, 1997). Young adult subjects were first asked to rate how much they liked or disliked rock music. The subjects then listened to 10-second samples of rock music presented binaurally. There were nine samples ranging in amplitude from 10 dB to 90 dB above the individual subject's hearing threshold. The samples were presented in random order, and the subjects were instructed to rank the intensity of each sample by assigning a number. Fucci et al. (1994) asked their subjects to estimate the "loudness" of the music. They reported that the perceived loudness of the rock music was directly related to the females' musical preference. Females reporting that they "definitely dislike" rock music rated the music louder than females reporting that they "definitely like" rock music. Fucci et al. (1997) repeated the experiment with only one procedural difference. Instead of rating "loudness," the subjects were asked to rate "annoyance." This time it was the males who showed the greatest effect. The males who had reported that they "definitely dislike" rock music rated the music as more annoying than males who had reported that they "definitely like" rock music.

The senses of smell and taste

A number of studies have looked for differences in olfactory sensitivity between females and males. The results of these studies have been mixed. Although it is often reported that females are generally superior to males in olfactory function, the literature does not consistently support this view. As well as a generally greater sensitivity in females, some have reported differences in sensitivity to particular odors, while other studies have reported no differences for the same stimuli. Methodological problems may account for many of the discrepancies, particularly in the early studies. For example, odors are particularly difficult stimuli to deliver and remove; they diffuse easily and can remain for long periods. Or, if your subjects smoke or have recently been in the company of smokers, their sensitivity may well be altered. Koelega (1994) conducted a study of olfactory sensitivity designed to overcome some of the design problems of previous studies. One hundred and twelve subjects with a mean age of 20 years participated in the experiment. There were 57 females (30 nonsmokers and 27 smokers) and 55 males (26 nonsmokers and 29 smokers). Five odorants were used: amyl acetate (banana oil), n-butanol iso-valeric acid (pungent, unpleasant), pentadecanolide (musky, scent of angelica root oil), and oxahexadecanolide (exaltolide, also musky). Subjects were presented with a series of four odorants in small glass bottles and were asked to indicate which odor was different from the other three. After a 60-second interval, the next array of bottles was presented. There were 4 test sessions separated by 24 hours, with 7 trials per session. Analysis of the results revealed two sex differences: females were more sensitive than males to musk and banana oil. Although there was a trend toward decreased sensitivity in smokers, the difference was not significant.

Olfactory event-related potentials (OEPs) are, as the name suggests, evoked potentials, in response to a particular odor. Olfactory acuity and OEPs were recorded from 33 subjects ranging in age from 18 to 83 years (Evans et al. 1995). Eighteen subjects were females; 15 subjects were males. The olfactory stimulus, amyl acetate (banana oil), was presented to the right nostril in different concentrations. Although the ability to detect the odor diminished with increasing age, there were no female-male differences in sensitivity. There were, however, significant differences in the N1-P2 amplitudes between females and males. The amplitudes for the female subjects were 60 to 90 percent larger than the amplitudes for the male subjects. A study using 40 odorants, the University of Pennsylvania Smell Identification Test, also demonstrated that olfactory sensitivity decreases with increasing age (Ship and Weiffenbach 1993). However, this study of 221 males and 166 females, aged between 19 and 95 years, also found a sex difference in sensitivity, with females demonstrating greater sensitivity than males. From as early as 1935, olfactory sensitivity has also been reported to vary across the menstrual cycle. It has been reported that in the pre-ovulatory phase, females are more sensitive to musk and rose. Premenstrual females have been reported to be more sensitive to benzene, coffee, and camphor (Figure 5.5; see Walker 1997 for a review).

During pregnancy, olfactory perception can change radically. The popular press and anecdotal evidence tells us that previously preferred scents can become repugnant and previously neutral scents can elicit bouts of vomiting and morning sickness. A study of 126 pregnant and 76 nonpregnant females matched for age and smoking habits were enrolled in a longitudinal study. The pregnant subjects answered questionnaires on smell and taste during gestational weeks 13–16 and 31–34, and 9–12 weeks postpartum. The nonpregnant subjects filled in questionnaires at similar time intervals. The number of pregnant subjects reporting abnormal smell sensitivity was eighty-five (68 percent) for weeks 13–16, forty (33 percent) for

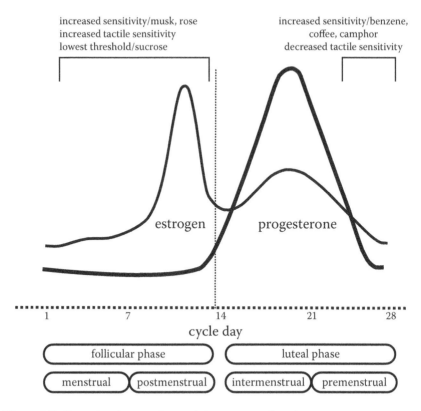

Figure 5.5. Sensation/perception and the menstrual cycle.

weeks 31–34, and four (3 percent) for the postpartum questionnaire. The reports by the nonpregnant females were five (7 percent), two (3 percent), and zero for the same time periods. In relation to abnormal taste sensitivity, the values for the pregnant group were thirty-three (26 percent), sixteen (13 percent), and four (3 percent). The corresponding values for the nonpregnant group were six (8 percent), zero, and one (1 percent) (Nordin et al. 2004). One of the shortcomings of this type of self-report study is that although the subjects experienced altered sensations, no objective measures of those sensory experiences were made.

This problem was overcome in a later study (Cameron 2007) in which objective measures of olfactory sensitivity were correlated with self-reports of olfactory sensation in subjects who were pregnant (first trimester, n = 20; second trimester, n = 20; third trimester, n = 20); postpartum, within 3 months of delivery (n = 20); or not pregnant (n = 20). The experimental procedure used the 40-item University of Pennsylvania Smell Identification Test. For each item, the participants were required to make a forced choice identification between four alternatives by scratching odor strips to release the individual smells. In addition, for each item the participants were asked to make a verbal rating on a seven-point scale of the intensity and pleasantness of each odor. The experimenters also counted the number of times that the subjects scratched and sniffed each odor strip as an indicator of sensitivity to the odors. Before commencing the testing, each participant was also asked to rate the sensitivity of her sense of smell. Sixty-one percent of the pregnant subjects rated their sense of smell as having become more sensitive during pregnancy. This is particularly interesting as, despite the self-rating, there was no significant difference in odor sensitivity between any of the groups of subjects, although subjects in the first trimester needed fewer scratches of the odor strip in order to make an identification. Using Chi-square analysis of the individual ratings, three odors (leather, lemon, and natural gas) were rated significantly more intense and three odors (orange, grape, and natural gas) were rated as significantly less pleasant when first-trimester responses were compared with control responses. The authors concluded that although self-reports from 61 percent of pregnant subjects claimed greater sensitivity to odors, this perception was not borne out by the experimental results. Interestingly, the pregnant brain, particularly in the first trimester, seemed to be altering the interpretation of the consistent sensory data.

Although there is a lot of anecdotal evidence suggesting that cravings for sweets, and chocolate in particular, occur more frequently in the premenstrual phase, little work has been done on taste preference and the menstrual cycle. The threshold to detect sucrose has been reported to be lower in the pre-ovulation phase (Than et al. 1994). A premenstrual increase in the intake of high-fat, high-carbohydrate foods has also been reported (Rogers and Jas 1994). Finally, and this seems like a good place to leave the discussion of taste, it has been reported that females prefer the taste of "expensive" brands of chocolate-chip ice cream, while males prefer the taste of the cheaper brands (Kunz 1993).

Before leaving this section, a final word on the factor modulating food intake, hunger. Hunger and eating behavior is an area fraught with confounding factors. Young women may feel intense hunger but barely eat in an effort to control weight. Young men may eat far beyond hunger in an effort to increase their body size. A PET study of 22 females (in the follicular phase of their cycle) and 22 males has reported different patterns of brain activation during hunger. Males had significantly higher levels of activation in the frontotemporal and paralimbic areas than females. In response to satiation, on the other hand, females had greater activation in the dorsolateral prefrontal cortex and the occipital and parietal sensory association areas than males (Del Parigi et al. 2002). It has also been demonstrated that, at least under test conditions, males are better able to regulate their energy intake to their energy needs than females. A possible explanation, the authors suggest, for the greater predisposition for females to gain weight over time (Davy et al. 2007).

A word about touch

Tactile sensitivity is not an area that has been greatly researched. The threshold to detect vibrations delivered to the palm of the hand has been reported to be the same in females and males, although females report vibrations above the threshold as "more intense" than males (Verrillo 1979). A study applying a tactile stimulus (also vibration) to the tongue, failed to find a similar difference in judgments of above-threshold stimuli (Petrosino et al. 1988). It has also been reported that tactile sensitivity is greatest in the pre-ovulatory phase and lowest premenstrually (Walker 1997).

Numbers and sums

The ability to manipulate numbers is frequently used as a general measure of cognitive function. This ability may be tested in a variety of ways, but one of the more popular methods (particularly for screening job applicants) is some form of the "serial addition" task. This task requires the person to add one number to another, speak or write the answer, add a number to that answer, speak or write the new answer, and so on. Usually the process is timed or paced and both speed and accuracy are measured. The test requires more cognitive skills than just simple addition. It is necessary to do the arithmetic, make the response, listen for the next number to be added to the last answer, and not confuse the numbers in the present addition with those in any of the previous sums. Furthermore, a timer is ticking away or the new numbers to be added are simply arriving at a fixed rate, for example, one every two seconds.

Wiens et al. (1997) used the Paced Auditory Serial Addition Test (PASAT) to measure the performance of 821 adults (672 males and 149 females). The PASAT uses a prerecorded set of 50 single-digit numbers presented in random order. The task of the subject is to add each new number to the previous total and speak the new total. There are four trials, each with a different set of randomized numbers; the only difference between the trials is that the rate at which the numbers are presented increases across trials (2.4, 2.0, 1.6, 1.2 seconds between numbers, respectively). The variable measured is number of correct responses. In this study, the performance of females and males was equal for the first three trials; however, on trial 4 (one number every 1.2 seconds), the accuracy of the males was significantly greater than the accuracy of the females.

Another test of cognitive function that uses number skills is the "Mental Dice Task." This task requires the subject to produce a series of "random" numbers using numbers from 1 to 6 only, "as if throwing dice." As with serial addition, the task requires more than just number manipulation. To generate the response, the subject must overcome the natural tendency to count, either forward or backward, or to repeat a pattern of presentation, for example, 3 odd, 3 even. One particular advantage of this task is that the subjects do not know what response is expected of them. There is no "correct" answer. There is, however, a similarity with the Stroop test. In order to respond appropriately, the subject must suppress an "automatic" response, in this case counting. Brugger et al. (1993) tested 20 right-handed females (mean age, 30 years) during the preovulatory phase (days 8–13) and during the premenstrual phase (days 23–28). The subjects were instructed to imagine that they were rolling dice and to call out their imaginary numbers at a rate of one number per second. A series of 66 numbers was recorded in this way. When the numbers generated by the subjects were compared with a real series of dice throws, there was evidence of too much counting and, interestingly, too few repetitions at both test times. When comparing the responses between the two phases of the menstrual cycle, there was significantly greater counting in the premenstrual phase than in the preovulatory phase. If you consider this result as indicating a reduced suppression of an automatic behavior, then the result is consistent with the Stroop results of Lord and Taylor (1991).

Memory

It is difficult to talk about memory as a separate entity. All of the types of cognitive function covered so far in this chapter rest on a firm foundation of memory. The saliency of taste and smell relies in large part on memories of previous events of those tastes and smells. The sense of smell is a particularly potent reminder of things past. Even in old age the scent of a perfume or flower can bring back memories of a long-forgotten childhood. Colors are intrinsically linked with the memory of previous encounters with a particular hue. It is difficult for most people to remember color out of context: red is usually remembered as apples and fire trucks; green evokes leaves and grass. Learned and remembered skills are required for the successful completion of some tasks (mental rotation, giving directions) or need to be suppressed for the successful completion of others (Stroop, mental dice). What distinguishes the papers covered in the following section from those discussed above is their focus on the memory processes required for the cognitive task, rather than task performance per se.

A good place to start this discussion is with a return to the question of the hunters and the gatherers. Males are better at mental rotation of objects but women are better at remembering where hidden objects may be found. What is it about the object location task that gives females the advantage? James and Kimura (1997) approached this question by presenting subjects with a visual array composed of drawings of 27 objects. After studying the original stimulus array, the task was to identify objects in a test array which had exchanged location with another object (condition 1) (Figure 5.6) or moved to a new location without exchange (condition 2). The stimuli were presented either on printed pages or on a computer screen. The experimental subjects were 174 undergraduate psychology students aged between 18 and 27 years. Forty-one females and 43 males participated in condition 1; 46 females and 44 males participated in condition 2. The protocol was the same for the two conditions. The subjects were asked to study an array of objects. After 1 minute, the original array was replaced with a new, test array. The subjects were asked to place an "X" on objects that had moved and to circle objects that had not moved (Figure 5.6). For the subjects responding to the computer array, an

Figure 5.6 Moved objects. Subjects are required to study the stimulus array (a). The task is to identify the two moved objects in the test array (b).

overhead transparency was placed over the screen and they marked their responses on this. When the results were analyzed, there was no difference between paper presentations and computer presentations. In condition 1, location exchange, females performed more accurately than males. In condition 2, where objects had been moved rather than exchanged, female performance was similar to condition 1, and there was no difference between females and males. One of the major differences between the two conditions was in terms of filled and unfilled space. In condition 1, the pattern of filled and unfilled space remained the same. Two objects were exchanged but the occupied areas of the array remained constant. The critical feature was identification of the individual objects. In condition 2, the pattern of filled and unfilled space changed. An object was moved from its original location to a new place in the array, one gap was filled and another gap opened up. Identification of individual objects was de-emphasized. In terms of evolutionary advantage, James and Kimura suggested that for gatherers, correct object identification (edible versus nonedible) is the crucial discrimination. It is the individual objects that are remembered. For the wider-ranging hunters, however, recognition of filled and unfilled space (a familiar or an unfamiliar horizon) may have literally made the difference between life and death. So, in condition 2 where the males had the additional cue of the pattern of filled space, their performance was equal to the performance of females.

The kinds of objects in a visual array, female-oriented or male-oriented, have been suggested to be important to visuospatial memory. McGivern et al. (1997) tested the effect of object type on people's ability to remember images. The subjects in the study were 39 females (mean age 21 years) and 23 male undergraduates (mean age 22 years). The subjects were asked to study a page containing an array of items for 60 seconds. The array could be female-oriented (e.g., doll, dancer, dress), male-oriented (e.g., football, lion, rocket), or random (e.g., umbrella, phone, guitar). After studying the array, subjects were then given a page containing the original stimulus array, with additional items added. The task of the subjects was to circle those items not included in the original array. The female subjects' recall was better than the male subjects' recall for the random and female items. Female and male recall were equal for the male items.

It has been suggested that females and males may differ in the way they process and, therefore, remember a visual array. A study of 24 females and 24 males, aged between 18 and 48 years, assessed the subjects on four different aspects of their visuospatial working memory formation processes including image generation, maintenance, scanning, and transformation (Loring-Meier and Halpern 1999). For the first task, image generation, the subjects were asked to study an image of a letter, draw that letter from memory, and then decide whether an "X" on a computer screen would have been covered by the original (remembered) image. For the second task, image maintenance, the subjects were asked to memorize a shape on the computer screen, then decide if an "X" that appeared on the screen after the first image disappeared would have covered that image. The third task, image scanning, required the subjects to study an arrangement of black and white boxes on a computer screen. When the boxes disappeared, an arrow was displayed and the subjects had to decide whether or not the arrow pointed to the position of a black box in the previous array. The fourth task was a standard mental rotation task. The results of the study were surprisingly uncomplicated: males performed the tasks faster than females, but the accuracy of the performance was equal for females and males.

It has been reported that females perform better than males on tests of episodic memory, the memory of specific events or items. One suggestion to explain this difference is that females' superior ability in verbal tasks may generalize to give better performance on tasks testing episodic memory, for example, word-recall tasks. Herlitz et al. (1999) addressed this question by assessing a group of 20- to 40-year-olds (100 females, 100 males) on a series of nine tasks that measured episodic memory on both verbal and spatial tasks (Table 5.1).

In analyzing their data, the authors not only looked for male–female differences (male or female "advantage") on the different tasks, they also looked for relationships between verbal and spatial abilities and episodic memory. On the verbal tests, females performed better than males on the fluency and synonym tasks, but male and female performance was equal on the anagram task. As expected, males performed better on the mental rotation task, but there was no difference between females and males on the water-level task. On the episodic memory tests, females were more accurate than males on free recall for abstract words and concrete pictures. Male and female performance did not differ significantly for the other tasks. After this initial analysis, Herlitz et al. used factor analysis to tease out the possible relationships between the different tasks. Factor analysis is a form of statistical analysis related to matrix algebra. The idea is that data from a number of different tests can be analyzed, specifically looking for sets of variables whose individual variables initially correlate across subjects. Such sets of correlated variables are called factors. Three factors were found for the present study. The first factor grouped abstract word recognition (no advantage), abstract word recall (female advantage), concrete word recognition (no advantage), and concrete word recall (no advantage). The second factor contained synonyms (female advantage), fluency (female advantage), and concrete pictures free recall (also female advantage). This grouping strongly suggests that the female advantage for synonyms and fluency has contributed to the female advantage for the episodic memory task, concrete pictures free recall. The third factor

Table 5.1 Design of study by Herlitz et al. 1999.

Test	Task and Response Required	Variable Measured
Verbal:		
Fluency	Generate and write words beginning with "F," "A," and "S"	Number of words written in 1 minute
Synonyms	Generate and write synonyms for "modern," "aspect," "caring," "important," "rich," and "dull"	Number of correct synonyms written in 5 minutes
Anagrams	Solve 5-letter anagrams	Number of correct solutions in 5 minutes
Visuospatial:		
Mental rotation	Match target figures to rotated test figure	Number of correct figures in 10 minutes
Water level	Determine horizontal water level from view of tipped bottles	Number of correct water-level estimates; not timed
Episodic:		
Abstract words	View 30 abstract words (e.g., "statement," "feelings") for 3 sec.; free recall, write remembered words on paper (3 mins.); recognition test, indicate if viewed words were presented initially	Number of words recalled
Concrete words	View 30 concrete words (e.g., "window," "flower") for 3 sec.; free recall, write remembered words on paper (3 mins.); recognition test, indicate if viewed words were presented initially	Number of words recalled
Concrete pictures	View 30 concrete pictures (e.g., picture of window, picture of flower) for 3 sec.; free recall, write remembered pictures on paper (3 mins.); recognition test, indicate if viewed pictures were presented initially	Number of pictures recalled
Abstract pictures	View 30 abstract pictures for 3 sec.; recognition test, indicate if viewed pictures were presented initially	Number of pictures recalled

contained mental rotations (male advantage), abstract picture recognition (no advantage), and concrete picture recognition (no advantage). It is particularly interesting to note that even though males scored much better than females on mental rotation, they did not have an advantage on the other factor 3 tasks, abstract and concrete picture recognition.

In a study of verbal learning during pregnancy, 71 subjects who were 14 weeks pregnant and 57 closely matched nonpregnant control subjects were tested on a battery of verbal tests to assess verbal learning, semantic memory retrieval, and speed of information processing (de Groot et al. 2003). The speed of information processing between the two groups was the same; however, the pregnant group had lower (worse) scores on the tests of learning and retrieval from semantic memory. In a study of working memory and spatial ability during pregnancy, it has been reported that the sex of the fetus influenced the mother's performance. A male fetus stimulated his mother to outperform the mothers of a female fetus. This effect, the authors suggested, may account for the variability observed in cognitive tests conducted during pregnancy (Vanston and Watson 2005).

In a reversal of the usual behavioral testing tradition, a human analogue of the rat "radial arm maze" has been used to test spatial memory in females and males (Rahman et al. 2005). The performance of 26 females and 26 males, ranging in age from 18 to 45 years, was tested on the radial arm maze, mental rotation, and verbal IQ. For this task, 17 containers with detachable lids were placed in a circle on the top of a square table. Four identical chairs were positioned around the table. A number of other cues were placed around the room including a table, sideboard, picture, and clock. The position of the table and chairs was fixed relative to these cues. There were 10 small objects that could be placed in any of the 17 boxes. Following a familiarization period in which the subjects were asked to name and remember each test object, the main testing session began. Subjects were seated in one of the chairs and instructed to remember which four objects were used in the trial and which four of the 17 boxes had an object placed in it. They were specifically told that they did not have to remember which object was placed in which box. Between blocks of trials the subjects were instructed to move to different chairs, but the table arrangement remained the same. When the results of the tests were analyzed, females made fewer errors on the 17-box maze. When these results were further broken down, the female advantage was for the object memory component, leading the authors to suggest that this demonstrates a general female advantage in episodic memory. On the mental rotation task, the expected male advantage was found.

Female-male differences in event-related potentials (ERPs) during verbal and abstract figure memory tasks have been reported. A study of 24 females and 24 males has demonstrated that ERP latencies were longer for males than for females for both tasks (Taylor et al. 1990). The distributions of activity also differed between females and males, with females showing larger amplitude anterior P2 and N4 waves in response to abstract figures and larger amplitude posterior P2 and N4 waves in response to verbal stimuli. Auditory-evoked potentials, hypothesized to be related to acoustic memory, have also been reported to differ between females and males. Rojas et al. (1999) have reported that the time constant (the amount of time required for the response to decay) of the 100 msec auditory-evoked response is longer in females than in males. As the 100 msec response is suggested to underlie auditory sensory memory, this result may represent the biological basis of some of the observed behavioral differences.

Conclusions

In this chapter the results of quite a number of different studies have been presented. The problem, in trying to form the mass of divergent information into a coherent whole, is to decide what aspects of the studies are comparable and what aspects are not. The differences we are looking for may be small, on some tests half a standard deviation (Hampson 1990), and easily obscured. Hampson (1990) tackles this problem of comparability head-on. She designed a study to administer a battery of cognitive and motor tests to the same group of females, once during the menstrual phase and once during the mid-luteal phase. The tests included female advantage and male advantage tasks. The crucial point is that the tests were all administered under the same conditions, using the same equipment and the same protocols. Hampson's hypothesis was that during the mid-luteal phase when estrogen and progesterone are high, the performance on female advantage tasks should be at its best, while performance on male advantage tasks should be poor. Forty-five females, aged between 19 and 39 years (mean 24 years), were included in the study. The results of the study generally supported Hampson's hypothesis. During the menstrual phase, the subjects performed better on the male advantage tasks (e.g., spatial ability, deductive reasoning) than during the mid-luteal phase. During the mid-luteal phase, performance on female advantage tasks was better but the results were not as clear as the male advantage results (Hampson 1990).

The following summarizes the main points derived from this chapter. We will revisit some of these points in Chapter 7 in relation to psychiatric illness.

1. There are three categories of cognitive and perceptual tasks: female advantage tasks, male advantage tasks, and neutral tasks. The male advantage tasks are those requiring analysis and manipulation of spatial stimuli (e.g., rod and frame, mental rotation). The female advantage tasks are those requiring the analysis and manipulation of verbal and visual materials (e.g., synonyms, hidden objects).
2. Performance on male advantage tasks improves during the menstrual phase, when female performance becomes more like male performance. Performance on female advantage tasks is better in the mid-luteal phase.
3. The female-male differences on spatial tasks are related primarily to reaction time, while accuracy is usually equal. Females can perform the tasks as accurately as males, but it may take more trials to acquire the skill initially.
4. There is a female advantage for recognizing and remembering specific rather than global information.
5. There are differences in perceptual and sensory system sensitivity across the menstrual cycle. Prior to (or on) ovulation, visual acuity is greatest, sensitivity to musk is increased, and tactile sensitivity is heightened. This makes good sense since the systems most useful for finding a mate are most sensitive when the female needs them most (Ivory, unpublished observation).

Bibliography and recommended readings

Bibawi, D., B. Cherry, and J. B. Hellige. 1995. Fluctuations of perceptual asymmetry across time in women and men: Effects related to the menstrual cycle. *Neuropsychologia* 33: 131–38.

Brugger, P., A. Milicevic, M. Regard, and N. D. Cook. 1993. Random-number generation and the menstrual cycle: Preliminary evidence for a premenstrual alteration of frontal lobe functioning. *Perceptual and Motor Skills* 77: 915–21.

Butler, T., J. Imperato-McGinley, H. Pan, D. Voyer, J. Cordero, Y-S. Zhu, E. Stern, and D. Silbersweig. 2006. Sex differences in mental rotation: Top-down versus bottom-up processing. *NeuroImage* 32: 445–56.

Collins, D. W., and D. Kimura. 1997. A large sex difference on a two-dimensional mental rotation task. *Behavioral Neuroscience* 111: 845–49.

Crawley, R. 2002. Self-perception of cognitive changes during pregnancy and the early postpartum: Salience and attentional effects. *Applied Cognitive Psychology* 16: 617–33.

Dabbs, J. M. Jr., E-L. Chang, R. A. Strong, and R. Milun. 1998. Spatial ability, navigation strategy, and geographic knowledge among men and women. *Evolution and Human Behavior* 19: 89–98.

Davy, B. M., E. L. Van Walleghen, and J. S. Orr. 2007. Sex differences in acute energy intake regulation. *Appetite* 49: 141–47.

De Groot, R. H., G. Hornstra, N. Roozendaal, and J. Jolles. 2003. Memory perfor-
    mance, but not information processing speed, may be reduced during early
    pregnancy. *Journal of Clinical Experimental Neuropsychology* 25: 482–88.
Del Parigi, A., K. Chen, J-F. Gautier, A. D. Salbe, R. E. Prately., E. Ravussin, E.
    M. Reiman, and P. A. Tataranni. 2002. Sex differences in the human brain's
    response to hunger and satiation. *American Journal of Clinical Nutrition* 75:
    1017–22.
Desrocher, M. E., M. L. Smith, and M. J. Taylor. 1995. Stimulus and sex differences
    in performance of mental rotation: Evidence from event-related potentials.
    Brain and Cognition 28: 14–38.
Evans, W. J., L. Cui, and A. Starr. 1995. Olfactory event-related potentials in normal
    human subjects: Effects of age and gender. *Electroencephalography and Clinical
    Neurophysiology* 95: 293–301.
Feinman, S., and G. W. Gill. 1978. Sex differences in physical attractiveness prefer-
    ences. *Journal of Social Psychology* 105: 43–52.
Frost, P. 1994. Preference for darker faces in photographs at different phases of the
    menstrual cycle: Preliminary assessment of evidence for a hormonal rela-
    tionship. *Perceptual and Motor Skills* 79: 507–14.
Fucci, D., L. Petrosino, and M. Banks. 1994. Effects of gender and listeners' prefer-
    ence on magnitude-estimation scaling of rock music. *Perceptual and Motor
    Skills* 78: 1235–42.
Fucci, D., L. Petrosino, B. Hallowell, L. Andra, and C. Wilcox. 1997. Magnitude
    estimation scaling of annoyance in response to rock music: Effects of sex and
    listeners' preference. *Perceptual and Motor Skills* 84: 663–70.
Guiffre, G., L. DiRosa, and F. Fiorino. 2007. Changes in colour discrimination dur-
    ing the menstrual cycle. *Ophthalmologica* 221: 47–50.
Haiger, R. J., B. V. Siegel, A. MacLachlan, E. Soderling, S. Lottenberg, and M. S.
    Buchsbaum. 1992. Regional glucose metabolic changes after learning a com-
    plex visuospatial/motor task: A positron emission tomographic study. *Brain
    Research* 570: 134–43.
Hamel, C. J., and D. L. Ryan-Jones. 1997. Effect of visual detail on scene recogni-
    tion: Some unexpected sex differences. *Perceptual and Motor Skills* 84: 619–26.
Hampson, E. 1990. Variations in sex-related cognitive abilities across the menstrual
    cycle. *Brain and Cognition* 14: 26–43.
Herlitz, A., E. Airaksinen, and E. Nordstrom. 1999. Sex differences in episodic
    memory: The impact of verbal and visuospatial ability. *Neuropsychology* 13:
    590–97.
Hurlbert, A. C., Y. Ling. 2007. Biological components of sex differences in color
    preference. *Current Biology* 17: 623–25.
Ivory, S-J. Unpublished observation.
James, T. W., and D. Kimura. 1997. Sex differences in remembering the locations of
    objects in an array: Location-shifts versus location-exchanges. *Evolution and
    Human Behavior* 18: 155–63.
Johnston, V. S., and X-T. Wang. 1991. The relationship between menstrual phase
    and the P3 component of ERPs. *Psychophysiology* 28: 400–09.
Kim, S. H., and H. Tokura. 1997. Cloth color preference under the influence of
    menstrual cycle. *Applied Human Science* 16: 149–51.
Koelega, H. S. 1994. Sex differences in olfactory sensitivity and the problem of the
    generality of smell acuity. *Perceptual and Motor Skills* 78: 203–13.

Krug, R., R. Pietrowsky, H. L. Fehm, and J. Born. 1994. Selective influence of menstrual cycle on perception of stimuli with reproductive significance. *Psychosomatic Medicine* 56: 410–17.

Kunz, J. 1993. Ice cream preference: Gender differences in taste and quality. *Perceptual and Motor Skills* 77: 1097–98.

Lord, T., and K. Taylor. 1991. Monthly fluctuation in task concentration in female college students. *Perceptual and Motor Skills* 72: 435–39.

Loring-Meier, S., and D. F. Halpern. 1999. Sex differences in visuospatial working memory: Components of cognitive processing. *Psychonomic Bulletin & Review* 6: 464–71.

McBurney, D. H., S.J.C. Gaulin, T. Devineni, and C. Adams. 1997. Superior spatial memory of women; stronger evidence for the gathering hypothesis. *Evolution and Human Behavior* 18: 165–74.

McCloy, T. M., and J. M. Koonce. 1982. Sex as a moderator variable in the selection and training of persons for a skilled task. *Aviation, Space and Environmental Medicine* 53: 1170–72.

McGivern, R. F., J. P. Huston, D. Byrd, T. King, G. J. Siegle, and J. Reilly. 1997. Sex differences in visual recognition memory: Support for a sex-related difference in attention in adults and children. *Brain and Cognition* 34: 323–36.

McRoberts, G. W., and B. Sanders. 1992. Sex differences in performance and hemispheric organization for a nonverbal auditory task. *Perception and Psychophysics* 51: 118–22.

McWilliams, W., C. J. Hamilton, and S. J. Muncer. 1997. On mental rotation in three dimensions. *Perceptual and Motor Skills* 85: 297–98.

Mekarski, J. E., T.R.H. Cutmore, and W. Suboski. 1996. Gender differences during processing of the Stroop task. *Perceptual and Motor Skills* 83: 563–68.

Moody, M. S. 1997. Changes in scores on the mental rotations test during the menstrual cycle. *Perceptual and Motor Skills* 84: 955–61.

Neubauer, A. C., R. H. Grabner, A. Find, and C. Neuper. 2005. Intelligence and neural efficiency: Further evidence of the influence of task content and sex on the brain-IQ relationship. *Cognitive Brain Research* 25: 217–25.

Nordin, S., D. A. Broman, J. K. Olofsson, and M. Wulff. 2004. A longitudinal descriptive study of self-reported abnormal smell and taste perception in pregnant women. *Chemical Senses* 29: 391–402.

Penton-Voak, I. S., D. I. Perrett, D. L. Castles, T. Kobayashi, D. M. Burt, L. K. Murray, and R. Minamisawa. 1999. Menstrual cycle alters face preference. *Nature* (London) 399: 741–42.

Petrosino, L, D. Fucci, D. Harris, and E. Randolph-Tyler. 1988. Lingual vibrotactile/auditory magnitude estimation and cross-modal matching: Comparison of suprathreshold responses in men and women. *Perceptual and Motor Skills* 67: 291–300.

Rahman, Q., S. Abrahams, and F. Jussab. 2005. Sex differences in a human analogue of the Radial Arm Maze: The "17-Box Maze Test." *Brain and Cognition* 58: 312–17.

Richardson, J. T. E. 1994. Gender differences in mental rotation. *Perceptual and Motor Skills* 78: 435–48.

Rogers, P. J., and P. Jas. 1994. Menstrual cycle effects on mood, eating and food choice. *Appetite* 23: 289.

Rojas, D. C., P. Teale, J. Sheeder, and M. Reite. 1999. Sex differences in the refractory period of the 100 ms auditory evoked magnetic field. *NeuroReport* 10: 3321–25.

Sharps, M. J., J. L. Price, and J. K. Williams. 1994. Spatial cognition and gender: Instructional and stimulus influences on mental image rotation performance. *Psychology of Women Quarterly* 18: 413–25.

Ship, J. A., and J. M. Weiffenbach. 1993. Age, gender, medical treatment, and medication effects on smell identification. *Journal of Gerontology: Medical Sciences* 48: M26–M32.

Silverman, I., and K. Phillips. 1993. Effects of estrogen changes during the menstrual cycle on spatial performance. *Ethology and Sociobiology* 14: 257–70.

Silverman, J., M. Buchsbaum, and H. Stierlin. 1973. Sex differences in perceptual differentiation and stimulus intensity control. *Journal of Personality and Social Psychology* 25: 309–18.

Skinner, B. F. 1969. *Contingencies of Reinforcement: A Theoretical Analysis.* New York: Appleton-Century-Croft.

Taylor, M. J., M. L. Smith, and K. S. Iron. 1990. Event-related potential evidence of sex differences in verbal and nonverbal memory tasks. *Neuropsychologia* 28: 691–705.

Than, T. T., E. R. Delay, and M. E. Maier. 1994. Sucrose threshold variation during the menstrual cycle. *Physiology and Behavior* 56: 237–39.

Vanston, C. M., and N. V. Watson. 2005. Selective and persistent effect of foetal sex on cognition in pregnant women. *Neuroreport* 12: 779–82.

Verrillo, R. T. 1979. Comparison of vibrotactile threshold and suprathreshold responses in men and women. *Perception & Psychophysics* 26: 20–24.

Von Kluge, S. 1992. Trading accuracy for speed: Gender differences on a Stroop task under mild performance anxiety. *Perceptual and Motor Skills* 75: 651–57.

Walker, A. E. 1997. *The Menstrual Cycle.* London: Routledge.

Weekes, N. Y., and E. Zaidel. 1996. The effects of procedural variations on lateralized Stroop effects. *Brain and Cognition* 31: 308–30.

Wiens, A. N., K. H. Fuller, and J. R. Crossen. 1997. Paced auditory serial addition test: Adult norms and moderator variables. *Journal of Clinical and Experimental Neuropsychology* 19: 473–83.

chapter 6

Laterality

A good indicator of the impact that a scientific discovery has made on the general population is to apply the "gossip test." Does the discovery feature in the popular press? Do talk show hosts allude to it? Can you buy books telling you how to understand it, live with it, exploit it, enhance it, or make money from it? Using these criteria, asymmetry of brain function, "laterality," passes the gossip test.

The idea of "two" brains with different skills has captured the public imagination. Not surprisingly, it is the "creative" right brain that has attracted the most interest. A quick browse through an Internet bookstore will yield some interesting titles. Apparently, in addition to drawing with the right side of the brain, it is also extremely useful for cooking vegetarian food, riding horses, and coloring Tibetan patterns. In addition, if you are a "right-brain person," you can learn how to use left-brain financial skills, manage your career, and guide your right-brained children through the left-brained world.

Asymmetry in brain function has been recognized for over 100 years. It is not to be confused with the crossover in motor pathways leading from the brain. Each side of the brain represents the opposite side of the body. A stroke that damages the left hemisphere will result in paralysis or other motor deficits on the right side of the body, as well as speech disruption.

Much of the early evidence on laterality came from studies of head injury patients. Similar lesions, but on opposite sides of the brain in different patients, produced remarkably different effects. Lesions in discrete regions of the frontal and temporal lobes were demonstrated to produce profound but region-specific language deficits. It was initially thought that laterality was primarily important for language. Recently, this view has changed. It is now understood that the two hemispheres play different but complementary roles in analyzing, and responding, to a wide range of sensory stimuli.

With the discovery of the complementary roles that the two hemispheres play comes a better understanding of why we should have two hemispheres at all. When you think about the way our bodies are designed, it is not really clear why there is, for example, one heart but two kidneys, one liver but two lungs. A "spare parts" hypothesis has been suggested. If you lose one kidney there is another one in reserve, but livers are essential, too, and there's only one. The same applies to the heart and lungs. In thinking about the brain, the most accurate description is probably neither "one brain" nor "two brains," "one and a half" is most accurate. The diencephalon, brain stem, and cerebellum are not split. The division of structures and functions is restricted to the areas where the most sophisticated processing takes place, where consciousness is thought to reside. Viewed from this perspective, it makes sense to divide the most complicated tasks into smaller subtasks, distribute those subtasks to different areas, and then process them simultaneously for maximum efficiency.

Parallel processing and hemispheric advantage

The important feature of laterality is that the two sides of the brain process the same information for different attributes at the same time (parallel processing). This division of labor in processing is often referred to as "hemispheric advantage." Under normal circumstances, the results of the processing are communicated between the two hemispheres. Because the transfer of information is rapid, the result is a comprehensive analysis of information with the features extracted by the left and right hemispheres integrated into an "information package." This is an extremely efficient method of data analysis and undoubtedly confers an evolutionary advantage in terms of survival (Geschwind and Galaburda 1985). The advantages for language processing are just one aspect of divided processing, which is particularly convenient for a language-using species.

It is difficult, in a person with an intact brain, to analyze the function of the two hemispheres independently. For a number of years, based primarily on the results of lesion studies, it was assumed that the left hemisphere was the dominant, more important hemisphere. This assumption is understandable in view of the behavioral effects observed following left and right hemisphere lesions. Lesions on the left tend to produce obvious, easily observed, disruptions to language. Lesions to the right hemisphere produce much more subtle deficits. In the absence of highly specialized neuropsychological testing, the effects of right hemisphere lesions may go undetected. Over the years, a list of attributes for the two hemispheres, in addition to "more important–less important" judgments has accumulated (see Table 6.1). The most commonly suggested description is of a verbal

Table 6.1 Processing characteristics attributed to the left and right hemispheres.

Left Hemisphere	Right Hemisphere
Language	Visual-spatial representation
Analytic	Holistic
Verbal	Spatial
Amplifies small scale information	Amplifies larger-scale information
Categorical representations	Prosody
	Topographical properties

left hemisphere and a spatial right hemisphere. In truth, the distinctions are far more subtle and interesting!

Much of what we understand about laterality comes from studies of patients with lesions to particular brain regions. Interpretation of lesion studies is always difficult. You look at behavior following a lesion, try to determine how the behavior has changed from the pre-lesion behavior, then work backward to try to determine what the area was doing in the first place. So, for example, a patient may be shown a figure containing a capital letter composed of a number of small letters (see Figure 6.1a). After the figure has been hidden the patient may be asked to draw the figure from memory. The result may be something like the drawing in Figure 6.1b. The challenge is to determine, from that drawing, the nature of the functional changes that have occurred.

The passing of information between the two hemispheres, via the corpus callosum, is known as interhemispheric transfer. Interhemispheric transfer time (IHTT) can be measured using visual event-related potentials (ERPs), with the N1 component, measured at the parietal leads, being used to estimate IHTT. The IHTTs are not symmetrical: the IHTT from right to left is shorter than the IHTT from left to right (see Moes et al. 2007 for a review). Given the differences in the corpus callosum between females and males, it is logical to suggest that IHTTs might also be different. A study of 19 females and 19 males, all right-handed, with mean age of 19 years were included in a study of IHTT using ERPs. Based on previous studies, the researchers made three predictions for their results. First, they predicted that, as previously demonstrated, right to left transfer would be faster than left to right for all subjects; second, the overall the transfer times would be shorter for females than males; and third, that female IHTTs would be more symmetrical — that is, right to left would be similar to left to right. The subjects were seated at a computer screen with a central fixation point and instructed to watch the fixation point. Following a warning blink of the fixation point, two examples of the same letter were presented, either in the same or different case, for example, the letter A could be presented

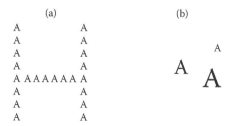

Figure 6.1 Global vs. local figure attributes. (a) The *global* letter, H, composed of the *local* letters, A. (b) Schematic representation of figure distortion following cerebral damage.

as "AA," "Aa," "aA," or "aa." The letters could be presented at two of four locations within a rectangle on the screen, for 60 milliseconds. The subject was instructed to make a "same or different" decision by pressing either the "M" key (match) or "N" key (no match) on the keyboard. The subjects' computer responses were recorded as well as a 16-electrode EEG. As the investigators predicted, IHTT was faster for right to left than left to right in all subjects. Females had significantly shorter IHTTs than males and a significantly smaller difference between the two transfer directions than males — that is, the responses in females were more symmetrical. One important limitation to this study was the use of only right-handed subjects. An earlier study using both left-handed and right-handed participants demonstrated significant effects of handedness, suggesting to the authors that left-handed subjects may have more efficient interhemispheric transfer of information (Cherbuin and Brinkman 2006).

Understanding language

Understanding and appreciation of language requires a number of different kinds of analysis. The temporal analysis of the auditory stimulus is followed by analysis of word patterns and speech rhythms, and, of course, meaning. The evidence is overwhelming that the two hemispheres make different contributions to this process. It is also clear that the processing differs between females and males.

By far the greatest volume of recent research on laterality is devoted to language. Interestingly, some of the research involves people yet to complete the language acquisition process. One particularly interesting study involved recording auditory-evoked responses (AERs) from 16-month-old infants as they listened to familiar and unfamiliar words (Molfese 1990). Eighteen infants took part in the study, nine females and nine males. The hand preferences of the parents were matched between the two groups. AERs were recorded from the infants while they listened to a tape-recorded voice reading a list of words, a repetition of two words, one word familiar to the infant and one word unfamiliar. In order to make the words as consistent as possible, the spoken words (female voice) were recorded and then edited to a 474-msec duration. A different tape was constructed for each infant, based upon interviews with the parents when words known and not known to the child were identified. During the testing sessions the parents were asked to observe their child's reaction to the word to confirm its "known" status. In addition, two observers also rated the child's reaction to each word. Analysis of the AERs showed that the response to unknown words was a larger amplitude negative peak than the response to known words, occurring at about 180 msec after the onset of the word. In the female infants this differential response was recorded at the left temporal electrode. In the male infants, however, differences for known and unknown words were recorded at the left frontal, left temporal, and right frontal electrodes. The meaning of the female-male response differences is not clear. The author suggests that the responses of the female infants may reflect either greater functional maturity or "a more lateralized system" (Molfese 1990, 359). This is interesting in view of the adult evidence to follow.

In 1995, clear sex differences in language processing were first demonstrated by Shaywitz et al., using MRI (Shaywitz et al. 1995; Pugh et al. 1996). Using echoplanar functional MRI, the brain activity of nineteen females (mean age 24.0 years) and nineteen males (28.5 years) was measured during four different language-related tasks. All of the participants were right-handed and all had reached the postgraduate level in education. The tasks all employed a visual display and the subjects were asked to make "same" or "different" judgments that were recorded by pressing a bulb to signal "same."

The first task required the subjects to view a display of two sets of four lines with different left-right tilts and to determine if the upper lines matched the lower lines in angle of tilt (Figure 6.2a). This task provided the control condition for the visual component of the following three tasks. The second test, the orthographic condition, required a same-different judgment for a series of consonants in upper- and lowercase letters (Figure 6.2b). In the third test, the phonological condition, the task was to decide whether two nonsense words rhymed (Figure 6.2c). In the final condition, the semantic category, subjects were asked to judge whether two real words came from the same semantic category (see Figure 6.2d). For analysis of the results, the first category, line orientation, was used to establish the baseline activity associated with the processing of the visual arrays.

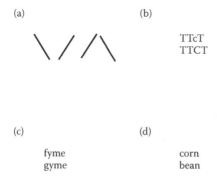

Figure 6.2 Schematic representation of the types of stimuli used in the "same-different" matching task by Shaywitz et al. (1995). (a) line recognition, (b) orthographic category, (c) phonological category, (d) semantic category.

Activity in seven different regions of interest (abbreviated, not surprisingly, "ROI") in each hemisphere was analyzed (see Figure 6.3). The accuracy in completing all four tasks was equal and high for both sexes, an average of 95 percent correct responses. An analysis of activity from all ROIs (total area analysis) showed that while overall activity increased significantly for males between the phonological condition and the semantic condition, a similar increase was not observed in the female subjects. In addition, left and right hemisphere activation was equal in females, but left hemisphere activity was greater than right hemisphere activity in males. Region-by-region analysis showed that in the occipital areas males showed increased activation for real words, compared with nonwords; however, females did not exhibit this increase. In the frontal regions, activation was bilaterally distributed in females but was lateralized to the left in males. The sex differences were most apparent in the inferior frontal gyrus where greater than 50 percent of the females showed bilateral activation. No males showed bilateral activation in this region. A similar pattern of activation was also seen in the orbital gyrus; this region usually shows the greatest activation related to phonological processing. In the temporal regions there was a significant increase in activation associated with semantic processing relative to the rhyme task, in males but not females. Based upon their results, the authors suggested that processing of reading may be divided into four anatomically distinct components: (1) visual processing, striate and extrastriate occipital areas; (2) orthographic representation, lateral and medial extrastriate areas; (3) phonological

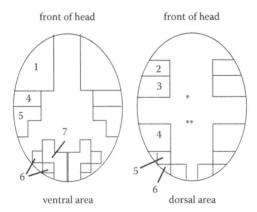

Figure 6.3 Schematic representation of the areas of interest of Shaywitz et al. (1995), Pugh et al. (1996). 1, lateral orbital gyrus; 2, prefrontal dorsolateral area; 3, inferior frontal gyrus; 4, superior temporal gyus; 5, middle temporal gyrus; 6, lateral extrastriate area; 7, medial striate area. Adapted from Pugh et al. (1996). For comparison with Frost et al. (1999), * represents thalamocapsular area, ** represents retrosplenial area.

representations, inferior frontal gyrus and temporal lobe (left hemisphere in males, bilaterally in females); and (4) lexical and semantic processing, middle and superior temporal gyri.

By contrast, a study published in 1999 by Frost et al. reported that language-related activity was clearly increased in the left hemisphere for both females and males during a language comprehension task. In this study, the subjects were 100 right-handed, native English speakers (50 female, 50 male), who were matched for age and education. The mean ages of the subjects were 24 years (males) and 22 years (females). A "semantic monitoring" experimental task required the subjects to listen to words naming animals (e.g., rabbit) and to respond, by pressing a button, only to those animals which were "found in the United States" and "used by humans." A "tone-monitoring" task, used to establish baseline activity for auditory stimulation, required the subjects to listen to a series of tones. The series could contain three to seven tones of high or low frequency. The subjects were instructed to respond only to a tone series that contained two high tones. The number of correct responses on both the tone-monitoring task and the semantic-monitoring task was the same for females and males. On the semantic task females scored a mean of 90.4 percent correct responses, while males scored 90.8 percent. On the tone task the female and male correct scores were 97.1 percent and 97.6 percent, respectively. Analysis of the activation patterns also showed little difference between females and males. The areas of greatest activation were in the left prefrontal, temporal, angular, retrosplenial, and thalamocapsular areas (see Figure 6.3). The only region showing greater right-sided activity was the cerebellum. Frost et al. concluded that it is questionable whether large sex differences in processing patterns exist.

In comparing the two studies, one of the most obvious differences is the difficulty of the linguistic aspects of the task. Pugh et al. required their subjects to perform tasks of increasing linguistic difficulty. The third condition in their study required the subjects to imagine the pronunciation of the nonsense words in order to determine if the words would rhyme when spoken. The fourth condition, the semantic category, required word recognition, then word comparison for meaning. Finally, the word categories had to be determined and the two words compared. By contrast, in the Frost et al. study, many of the linguistic characteristics were bypassed. The subjects knew the words would be animal names. The analysis required complex processing for categories and knowledge including subjects such as geography and animal husbandry. A much more extensive memory processing was required for this task. Given the differences in the experimental designs of the two experiments, a direct comparison of the results is not possible.

Jaeger et al. (1998) have suggested that it is the degree of difficulty of linguistic tasks that determines whether or not sex differences appear.

They suggest that at least some of the contradictions in the literature may be due to differences in the difficulty of the experimental tasks. In order to examine this possibility, they used PET to measure cerebral blood flow (CBF) while their subjects completed a series of language tasks. The subjects were eight females and nine males, all right-handed, native English speakers. In the baseline conditions, CBF was measured while the subjects fixated a visual target, "resting state." There were five tasks for the subjects to perform: (1) read aloud a series of verbs; (2) read aloud a series of nonsense verbs (e.g., "jelt," "brep"); (3) view then speak the past tense of regular verbs (e.g., "jump," "dance"); (4) view then speak the past tense of irregular verbs (e.g., "fall," "build"); and (5) view then speak the past tense of nonsense verbs not previously viewed. When the results were analyzed, it was clear that both reaction time and accuracy were equal for the males and females. However, differences in the pattern of CBF emerged that were dependent upon the task being performed. There was an overall increase in CBF in the bilateral occipital and motor areas, and the cerebellum. There was also an overall increase in the left frontal areas. These increases were the same in the females and the males. On tasks 1 and 2 (the easy tasks), there was a bilateral increase in the frontal and temporal areas that was equal in females and males. In the more difficult tasks (3, 4, and 5), the increased blood flow became lateralized to the left in males, but remained bilateral in females. When the CBF patterns of the two groups were compared with their respective baseline conditions, other differences became apparent. In all conditions, there was a greater increase in CBF in females bilaterally in the inferior occipital cortex and the cerebellum. On the tasks requiring past-tense generation, females also showed a significantly greater increase in the right anterior temporal and right posterior frontal cortices. These areas overlap the areas where Shaywitz et al. found greater activation in females during the phonological tasks. These results support the authors' suggestion that it is the difficulty of the linguistic task that distinguishes the female and male response patterns.

Anatomical differences have also been correlated with sex differences in cognitive performance. Gur et al. (1999) used MRI to study sex differences in the volume of gray and white matter of 80 right-handed people, 40 females and 40 males, aged between 18 and 45 years. The two groups were well matched on factors such as age, education, and IQ. In addition to the measures of tissue volume, the subjects were also tested using two measures of language ability and two tests of spatial ability. The results of this study are complicated. Readers who are interested in the details of the analysis should really consult the original paper. Briefly, females showed a higher percentage of gray matter than males, while males showed a greater percentage of white matter and CSF. But while the distribution of gray and white matter was symmetrical in females, in males the percentage of gray matter was greater in the left hemisphere, white matter was symmetrical and CSF was greater on the right. The test of cognitive function showed that overall, females and males performed equally well (the Global score). When the tasks were analyzed by verbal and spatial subcategories, however, females performed better on the language tasks than the spatial tasks, while males performed better on the spatial tasks. In both females and males, the Global score was correlated with brain volume. When analyzed separately, verbal performance was correlated with cranial volume in females. Performance on the spatial tasks was correlated with cranial volume for both females and males.

This study is an interesting return to the old question of brain size differences between females and males. Although the total brain volume in this study was smaller for females, the authors pointed out that the larger proportion of gray matter accompanied by a smaller proportion of white matter probably equalizes the differences. There is a larger percentage of tissue available for processing (greater percent gray matter) but shorter distances between brain regions (shorter axonal length, i.e., white matter, required).

A functional/anatomical study of magnetoencephalographic auditory evoked fields in response to an unattended auditory stimulus (pips) reported female-male differences in the localization of the source of the M100 wave. Seventeen females and males (one left-hander in each group) were tested. The source of the M100 wave was in the superior temporal gyri, located along the gyri of the left hemisphere in the female and male subjects. In the right hemisphere, the M100 source was in the same location in females but was found to be significantly further forward in males (Reite et al. 1995).

Because the hand preference of the subject is central to the question of laterality, many studies limit their experimental subjects to right-handed (or less frequently, left-handed) individuals. An interesting exception is a study by Okada et al. (1993), who used a measure of brain oxygenation to look for interactions between handedness and sex. Near-infrared (NIR) spectrophotometry was used to measure changes in blood oxygenation in the forebrains of 72 subjects while they performed a mirror-drawing task. In this task, the subjects must trace a figure, in this case a star, which they can see only indirectly via the reflection in a mirror (see Figure 6.4). The task is to trace the star as fast and accurately as possible. The 72 experimental subjects were subdivided into four groups: right-handed males (n = 22); left-(mixed) handed males (n = 20); right-handed females (n = 16); and left-(mixed) handed females (n = 14). The left-(mixed) classification was used because, as the authors point out, people in the age group of their subjects would have grown up during a time when left-handed children were trained to write and use chopsticks (this is a Japanese study) with their right hands. Three different patterns of blood oxygenation were observed. The "dominant hemisphere" response pattern was associated with an increase in blood volume in the dominant (as determined by

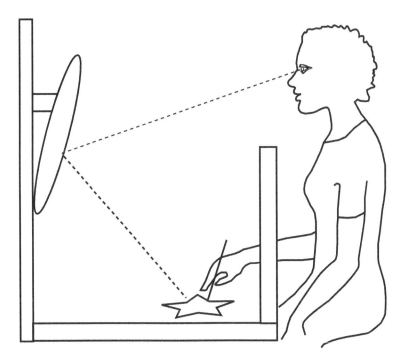

Figure 6.4 Schematic representation of the mirror-drawing task.

handedness) hemisphere. For example, a right-handed subject showed an increase in the left hemisphere. A "bilateral" response pattern was associated with a symmetrical increase in the two hemispheres. Finally, in two subjects, one male and one female, no change in blood flow was observed during the task, a "no response" pattern. The results are summarized in Table 6.2. From the results table, a different response pattern between females and males is easy to see. The majority of males in both groups exhibited a dominant hemisphere response pattern while the majority of females exhibited a bilateral response pattern. In fact, no left-handed female exhibited a dominant hemisphere pattern. In this respect, the greatest sex differences were found in the left-handed group. The results of this study suggest that for the kind of visual and motor processing required for the mirror-drawing task, females use bilateral processing while males process primarily with their dominant hemisphere.

Consider for a moment what these results would have looked like if this had been an experiment with only right-handed subjects. The results would have indicated left hemisphere processing in males (13/22 or 59 percent) and predominantly bilateral processing in females (11/16 or 69 percent). It is results such as these that illustrate the importance of experimental design.

Table 6.2 Results of the mirror-drawing task.

Hand preference	Dominant hemisphere response pattern		Bilateral response pattern		Total
	n	%	n	%	
Right-handed:					
Male	13	59	8	36	21
Female	4	25	11	69	15
Total	17	45	19	50	36*
Left-handed:					
Male	15	75	5	25	20
Female	0	0	14	100	14
Total	15	44	19	56	34

Source: Adapted from Okada et al. (1993).

* One female and one male showed no response pattern (not shown in table).

Emotion and laterality

The laterality studies discussed so far have dealt with processing of well-defined stimuli such as lines, tones, and words. Much more difficult to define are stimuli intended to carry an "emotional" component. The human face is often the stimulus selected to portray emotional content. In a study published in 1980, Ladavas et al. used photographs of actors' faces, expressing six different emotions, to measure the reaction time to discriminate different emotions.

This study was conducted at a time when computerized experiments were rare. The methods, by today's standards, seem primitive. The subjects were seated in front of a translucent screen and slides of the faces showing different expressions (happiness, sadness, surprise, fear, disgust, anger) were projected onto the back of the screen for a period of 150 msec. The subjects were required to fixate a central point on the screen. The photograph was projected on either the left or right portion of the screen. The photo on the right side of the screen was assumed to stimulate the left hemisphere; the left image was assumed to stimulate the right hemisphere. In each trial a number of photos were presented 7 seconds apart and the subject was instructed to press a response key as quickly as possible only when the target emotion for the trial (e.g., disgust) appeared. The reaction time (RT), the amount of time taken to press the button after the picture appeared, and the number of correct responses were recorded.

The responses of 24 right-handed subjects, 12 females and 12 males, were measured. Analysis of the results showed that, overall, RTs were faster for faces presented in the left visual field and that RTs in females were faster than RTs in males. The left-right difference in RTs was greater in females than in males. The accuracy in identification of the facial expression increased from trial 1 to trial 3, a good example of the practice effect. From this study, the authors concluded that the right hemisphere in females is superior to the left in discriminating emotional stimuli.

In 1983, a similar study was conducted using a larger number of subjects, 45 females and 45 males, divided into subgroups according to handedness (Natale et al. 1983). There were two categories of left-handers, those with an inverted writing posture (n = 15 females, 15 males) and non-inverted writing posture (n = 15 females, 15 males). A tachistoscope was used to view the individual stimulus cards. The duration of viewing time was adjusted according to individual thresholds for viewing, with a mean viewing time of 117.5 msec. The facial expressions portrayed were the same as the previous experiment. The response made by the subjects was different. In this case, the subjects were asked to adjust a lever to indicate how "sad" (1 = "very sad") or "happy" (10 = "very happy") the faces were. The female subjects gave a consistently "sadder" rating. When this result was analyzed further, it was found that the difference was due to the tendency of the female subjects to give a lower rating to the pictures displaying sadness, anger, disgust, and fear. There was no difference between females and males in their responses to the positive emotions. When the data were analyzed for handedness, the ratings of the left inverters were higher than the left non-inverters' and right-handers' scores. Interestingly, there was also a visual-field effect with faces presented to the left visual field receiving overall lower ratings. The right-handers had the largest visual field effect; the left inverters had a smaller effect. There was no visual field effect for the left non-inverters. When the visual field effects were further analyzed, they were found to apply only to the negative stimuli. From this result the authors concluded that the right hemisphere might be biased towards the emotionally negative aspects of a stimulus.

One hundred and four subjects took part in the second experiment. The groups consisted of right-handed males (n = 22), right-handed females (n = 16), left-handed inverter males (n = 19), left-handed inverter females (n = 8), left-handed non-inverter males (n = 19), and left-handed non-inverter females (n = 18). The stimuli were faces composed of two half-faces showing either happy or sad expressions. The two sides either matched, giving a whole "happy" or "sad" face, or mismatched, with one side "happy" and the other side "sad." The viewing time was much shorter than in the first experiment. The mean viewing time was 63 msec, just long enough to allow recognition that a face had been presented. As in experiment 1, females gave a lower overall rating than males, but there was no effect of handedness. Females appeared to be better at differentiating between the three face types. The subjects seemed to be better at discriminating "happy" from "sad" when the presentations were to the left visual field. Other experimenters have reported similar results (Bibawi et al. 1995).

Laterality, the menstrual cycle, and hemispheric advantage

A considerable deficiency in the studies discussed so far is the failure of the experimenters to consider the stage of the menstrual cycle of the female subjects. Weekes et al. (1999) directly addressed this problem, in the context of the "dual route" model of hemispheric specialization. The dual route model suggests that there are two routes for the processing of language, a lexical route attributed to the right hemisphere, and both lexical and nonlexical routes operating in the left hemisphere. According to the model, the lexical route proceeds in stages from letter recognition through to semantic processing to phonological analysis and speech production. The non-lexical route (to which the ability to read and articulate nonsense words is attributed) proceeds from a letter to sound conversion stage through to phonetic construction to the same final speech production stage as the lexical route. To date, there is a body of experimental evidence both supporting and refuting the dual route model. The aim of this study was to try to dissect out individual differences that may have contributed to the contradictions in the experimental literature.

The study was composed of three experiments. In all three experiments, the subjects were presented with a letter string that was a real word or a nonsense word. The stimulus presentation was lateralized to the left or right hemisphere, or was bilateral. The subject was asked to decide whether or not the stimulus was a real word.

In the first experiment, the subjects were 12 females and 12 males, all-right handed, native English speakers. In the second experiment there was a total of 41 subjects, all native English speakers: 14 female, 10 male right-handers; 9 female, 8 male left-handers. In experiment 3, the subjects were 32 females, 16 mid-luteal phase and 16 menstrual phase. The subjects' responses were analyzed for reaction time and number of correct responses, as well as the interactions of a number of variables including word length, frequency of presentation, word regularity, and hemisphere.

The first variable to consider is the frequency of word presentation. Responses to high-frequency words were more accurate than responses to low-frequency words, with females being more accurate than males. Also, responses to right visual field (RVF) presentations were more accurate than responses to left visual field (LVF) presentations. In males, the RVF advantage was significantly greater than in females. When considering cycle phase and word frequency effects, the menstrual phase group showed a greater RVF advantage for low-frequency words due to a significantly poorer performance for low-frequency words presented to the LVF. The responses of the mid-luteal group for RVF advantage was equal for high- and low-frequency words. On word regularity, with bilateral presentations, females were significantly less accurate on irregular words. Males were significantly worse on accurately identifying regular words. Females showed an RVF advantage for regular but not irregular words. Males showed a similar RVF advantage for both categories. In discussing their results, Weekes et al. concluded that, at least in their study, the right hemisphere seems to be more sensitive to individual differences, including sex differences. In this study, the RVF advantage for low-frequency words (as a result of poor right hemisphere performance) during the menstrual phase, disappeared with the mid-luteal estrogen peak (Figure 6.5). The authors suggest two possibilities to explain this effect: changes in callosal efficacy related to low estrogen; or a reduction in the

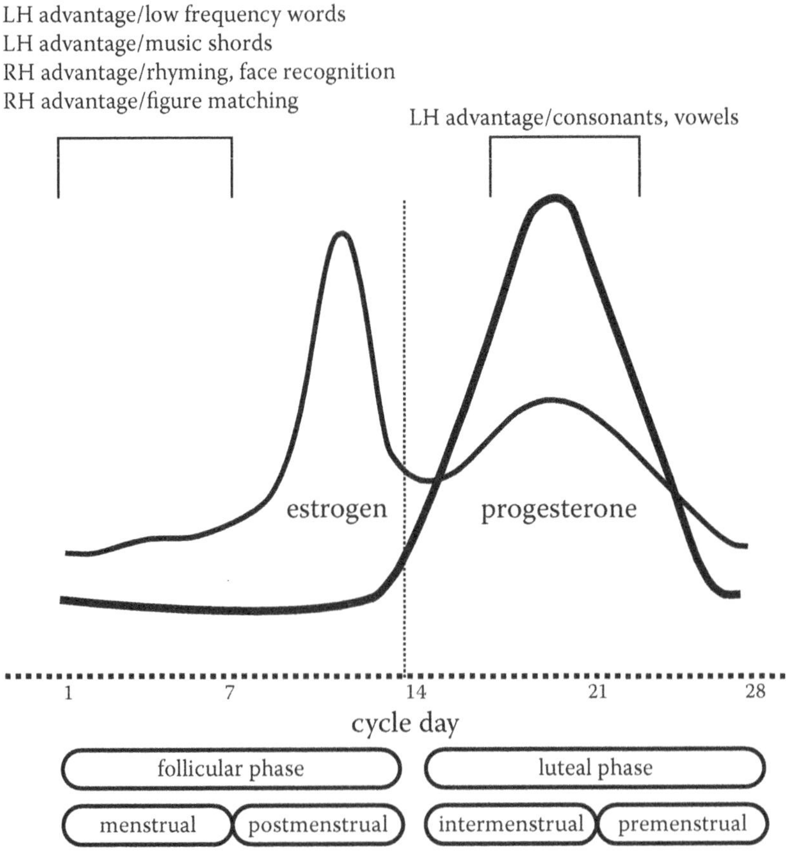

LH advantage/low frequency words
LH advantage/music shords
RH advantage/rhyming, face recognition
RH advantage/figure matching

LH advantage/consonants, vowels

estrogen   progesterone

1   7   14   21   28
cycle day

follicular phase   luteal phase

menstrual   postmenstrual   intermenstrual   premenstrual

Figure 6.5 Hemispheric advantage and the menstrual cycle.

right hemisphere's processing ability related to low estrogen. Based on these results, they favor the latter explanation.

Sanders and Wenmoth (1998) suggested that hemispheric advantage is the crucial factor in interpreting cyclic cognitive changes. Although many inconsistencies appear in the literature, the authors argued that when hemispheric advantage is considered, the inconsistencies are resolved. Specifically, the authors suggested that for tasks that generate a right hemisphere advantage (e.g., nonverbal tasks), the asymmetry is greatest when estrogen is low. For left hemisphere (e.g., verbal) tasks, the asymmetry should be greatest when estrogen is high.

Sanders and Wenmoth tested their hypothesis using a verbal task requiring consonant-vowel identification and a nonverbal, music chord recognition task. Thirty-two right-handed women participated in the study. The two tasks were administered during the menstrual phase and mid-luteal phase in a counterbalanced design. The stimuli were delivered to either the right or left ear and the number of correct responses was measured. For the vowel-consonant identification task, accuracy was greater for right ear presentation (left hemisphere) at mid-luteal phase than in the menstrual phase. For the music chord recognition task, accuracy was greatest for left ear presentations (right hemisphere), as would be expected for a nonverbal task, but accuracy was greater in the menstrual phase than during the mid-luteal phase. The changes between the menstrual and mid-luteal phase responses were due to a significant left ear (right hemisphere) performance decrement. There was only a small, nonsignificant improvement in left hemisphere performance at the same time. Other authors have reported similar results. In a study using a rhyming task and a face recognition task, the accuracy of identification of stimuli presented to the right hemisphere was greater during the menstrual phase for both kinds of stimuli (Mead and Hampson 1995) (Figure 6.5). Rode et al. (1995) reported that there was an LH advantage for a lexical decision task and that it did not change across the menstrual cycle. On a figure comparison task, reaction times were fastest and equal for LVF and RVF presentations during the luteal phase. However, during the menstrual phase, RT was greater for presentations to the RVF, indicating a right hemisphere advantage. This result is also generally consistent with the results of the Weekes et al. study previously discussed.

Two theoretical perspectives on laterality

Much of what we know about hemispheric function comes from perceptual studies, particularly from the area of visual perception. In *The Two Sides of Perception*, Richard Ivry and Lynn Robertson use the vast literature on perception and laterality to support the "Double Filtering by Frequency Theory" of hemispheric specialization that they propose.

In the areas of experimental psychology and visual perception, one of the most frequently asked questions is, "What do we see first?" When we are presented with a figure, for example, a letter composed of smaller letters (Figure 6.6), what do we recognize first? Do we perceive the large letter first (global analysis), then the small letters (local analysis), as we analyze for details? Or is it the reverse — do we first recognize the small letters that form the whole big letter? This question is extremely diffi-cult to test directly in people with intact brains. Accordingly, a number of cognitive psychologists are at odds with each other. Some propose a "down-up" order of processing where the local details are analyzed first, others propose "up-down" processing. In both cases serial processing is assumed. That is, one aspect of the figure is assumed to be analyzed before the other.

Ivry and Robertson, by necessity, took an indirect approach to the problem of global versus local features. "What," they asked, "happens to this type of cognitive processing in people who have suffered a brain lesion?" Ivry and Robertson analyzed the results of a number of visual perception experiments in which people with left- or right-sided lesions were asked to draw from memory a picture shown to them by the experi-menters. The results of their analysis were quite interesting. Individuals with lesions in the left hemisphere could draw the larger, global, aspects

```
       Z              Z
       Z              Z
       Z              Z
       Z              Z
       ZZZZZZZZZZ
       Z              Z
       Z              Z
       Z              Z
       Z              Z
       Z              Z
```

Test figure

Right hemisphere damage   Left hemisphere damage

Figure 6.6 Schematic representation of visual processing deficits associated with left and right hemisphere damage. Adapted from Delis et al. (1986).

of the pictures they had been shown, but were unable to reproduce the finer (local) details. Individuals with lesions in the right hemisphere could reproduce the fine, local, features of the picture, but were unable to organize these local features into a coherent, global image (see Figure 6.6). In the terms of visual perceptionists, the left hemispheres of the experimental subjects were analyzing the pictures for high-frequency information, and the right hemispheres were analyzing for low-frequency information.

Frequency analysis is an important tool for understanding the workings of the brain. Many parts of the CNS work in a frequency-dependent manner. For most people, frequency analysis for auditory stimuli is commonplace. Judgments about the "pitch" of a particular tone are a form of frequency analysis. High-frequency sounds, where the sound waves are close together, are perceived as high pitch. Low-frequency sounds, where the sound waves are farther apart, are perceived as low pitch. A healthy 20-year-old is able to hear sounds in a range from 20,000 Hz right down to 20 Hz. As we age, the hearing in the upper frequency range decreases gradually. This age-related decrease is not important for everyday life, since speech sounds are in the 600 to 2,000 Hz range (Figure 6.7). Visual information is analyzed in exactly the same frequency-dependent manner. Large items in the visual field are low-frequency stimuli, while the small details that make up the larger items are high-frequency stimuli. In visual perception experiments, ratings of high and low frequency (Figure 6.8), without the distractions contained in a picture, are used to study the responses of the visual system.

Having determined that the left and right hemispheres seemed to be analyzing visual information for high- and low-frequency components, respectively, Ivry and Robertson turned to the literature on auditory perception. If one of the main features of hemispheric processing was a frequency bias, they reasoned that you might expect such a bias to be observed for other sensory systems. Indeed, from their analysis of the literature, evidence for such a bias in the auditory system emerged. One factor, which complicates frequency analysis by the auditory system, is

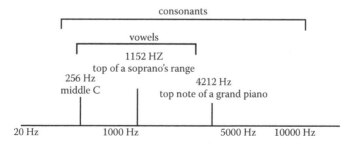

Figure 6.7 Frequency of selected sounds in the range of normal human hearing.

high spatial frequency   low spatial frequency

Figure 6.8 Examples of lines of relatively high and low spatial frequency.

the sequential nature of auditory stimuli. In the visual system, all of the parts of a picture are presented to the visual system at the same time. Complex auditory stimuli, on the other hand, are presented over time. For example, consider a spoken sentence. Single sounds, the vowels and consonants, are arranged into words, which are arranged to form the sentence. Frequency analysis of the sentence will include the frequencies of the individual sounds (local components) and the prosody or organization of words (global components). Interestingly, it appears that prosody is analyzed by the right hemisphere.

The examples above only touch upon the kind of evidence Ivry and Robertson used to synthesize the "Double Filtering by Frequency (DFF) Theory." A "bare bones" description of the theory follows. For readers who would like more detail, *The Two Sides of Perception* (1998) is highly recommended.

There are three fundamental tenets of DFF:

1. Visual and auditory stimuli are analyzed on the basis of frequency. As discussed above, there is a great deal of experimental evidence to support this.
2. Processing of material is divided between the two hemispheres. Both hemispheres are able to process the full stimulus array, but the results of that processing will be represented differently by each hemisphere. In addition, the two hemispheres will share their processing results via the corpus callosum, making it difficult to distinguish which side is doing which part of the analysis.
3. Selective attention is necessary for the frequency-based analysis to occur (and therefore, for hemispheric differences to become evident). There is evidence that the hemispheric differences do not occur if the individual is not attending to the properties of the particular stimulus.

In terms of describing processing attributes, the right hemisphere processes for and amplifies the topographical properties of the stimulus. In a visual display, this would include judgments on the distance between two large objects; in auditory terms, the prosody of the sentence. The left hemisphere processes for and magnifies the information on the fine details of the stimulus. It attends to the single letters on a keyboard or the vowels and consonants in speech.

If DDF provides a description of "what" happens, Miller's "Conduction Delay Hypothesis of Cerebral Lateralization" (CDH) tries to explain "how" it happens. It is clear that the two hemispheres process information differently and that they seem to be particularly adept at certain kinds of tasks. The question Miller tackles is, "What neurobiological processes underlie these observed differences?" For example, when a person listens to and comprehends spoken language, the information to be processed arrives not in a single presentation, as in a visual array, but as a series of sounds in a particular temporal sequence. The temporal nature of the speech requires a processing system with built-in delays, so that the first sound is related to the last sound in the sequence. The size of the necessary delays is tiny, in the range of 50 to 200 msec. In terms of synaptic transmission alone, however, that is a very long time to delay a signal. So the question is, how can this temporal analysis occur within the constraints of synaptic transmission? More specifically, what is different about the left and the right hemispheres that gives them their respective advantages in signal processing?

Drawing upon a wide range of experimental evidence from neuroanatomy, neurophysiology, and experimental psychology, Miller proposes a functional/structural basis for hemispheric lateralization. As in the case of the DDF, what follows is a "bare bones" summary and for the interested reader, *Axonal Conduction Time and Human Cerebral Laterality: A Psychobiological Theory* (1996) is recommended reading.

There are two ways that delays of 50 to 200 msec can be introduced into a neural pathway. First, by summing the delays produced by a large number of individual synapses. The incoming signal would have to be routed along multisynaptic pathways. At each synapse a degree of variability could be introduced which, when summed across a large number of synapses, would probably result in an inconsistent signal representation. Second, the delay could be introduced by slowing the conduction velocity along the axons themselves. If the signals were carried by small, unmyelinated axons, an appropriate delay could be introduced. In addition, the delay would be consistent across time. The conduction velocity of axons does not change under normal circumstances, and inconsistencies introduced by synaptic transmission would be minimized. Within their defined range, speech patterns should be consistent, as indeed they are. It is this second idea that underlies the CDH.

Stated simply, the CDH proposes that the differences in processing capabilities between the left and the right hemispheres are due to different proportions of "fast" and "slow" conducting axons. The nice thing about the CDH is that it generates a number of anatomical predictions that can be tested empirically.

Some of the major anatomical predictions are as follows:

1. The volume ratio of white to gray matter should be lower on the left than on the right. *There is a limited amount of data and the results are inconclusive.*
2. Cortical cell density should be greater on the left than on the right. *An early-twentieth-century postmortem study offers some support.*
3. The overall volume of the right hemisphere should be greater than the overall volume of the left. *There is evidence to support this prediction.* Miller concludes that the weight of evidence suggests larger frontal and temporal lobes on the right and larger occipital lobes on the left (at least for right-handed individuals).
4. The concentration of markers for myelin or myelin basic protein should be higher on the right than on the left. *There is no data at present on this.*

The support for these predictions is currently quite limited but nevertheless convincing in its consistency (see Im et al. 2008). In most cases, the number of studies is small. In addition, a variety of methods have been used, making interpretation difficult and comparison between the studies almost impossible. So, for example, for prediction 1 there are ten studies, conducted between 1914 and 1991 (Miller, p. 34, Table 3). The methods used in the different studies include volume estimation from cortical surface area, CT density, clearance of Xenon, and histological sectioning of postmortem tissue. In addition, an equal number of the studies have reported statistically significant and non-significant results, respectively.

Conclusions

Neuroscience is one of the most rapidly expanding areas in the biological sciences. It is hoped that within the next few years, experiments will be performed that will provide evidence to test Miller's predictions. Better yet, perhaps someone will set out to directly test the CDH. Either way, in the spirit of classic scientific theorizing, the predictions are there to be tested.

From the experimental evidence discussed in this chapter it appears that the sex differences reported in laterality studies are real. They are not straightforward, however. The following points summarize the main points derived from this chapter:

1. The patterns of activity associated with some aspects of language processing are clearly different in females and males. In females the tendency is for bilateral activation and reduced interhemispheric transfer times, with left-lateralized activation the prominent pattern in males.
2. Consideration of handedness is important for sex differences, at least for visuo-motor tasks. Processing in males increases in the dominant hemisphere, but females show a bilateral increase, regardless of handedness.
3. The right hemisphere in both males and females appears to be biased toward negative emotional stimuli. The right hemisphere's ability to discriminate emotional stimuli also appears to be better in females.
4. Hemispheric advantage is an important factor when considering changes associated with the menstrual cycle. For tasks with a right hemisphere advantage, the left-right asymmetry is greatest when estrogen is low. For tasks with a left hemisphere advantage, asymmetry is greatest when estrogen is high. Stated another way, right hemisphere ability is reduced when estrogen is high (Figure 6.9).

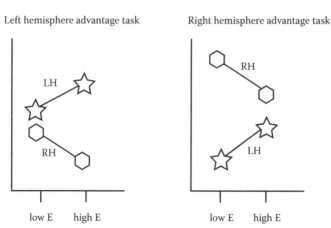

Figure 6.9 Schematic representation of the interaction between estrogen levels and task-associated hemispheric advantage. ☆ represents left hemisphere task performance; ○ represents right hemisphere task performance.

Bibliography and recommended readings

Bibawi, D., B. Cherry, and J. B. Hellige. 1995. Fluctuations of perceptual asymmetry across time in women and men: Effects related to the menstrual cycle. *Neuropsychologia* 33: 131–38.

Cherbuin, N., and C. Brinkman. 2006. Hemispheric interactions are different in left-handed individuals. *Neuropsychology* 20: 700–07.

Delis, D. C., L. C. Robertson, and R. Efron. 1986. Hemispheric specialization of memory for visual hierarchical stimuli. *Neuropsychologia* 24: 205–14.

Frost, J. A., J. R. Binder, J. A. Springer, T. A. Hammeke, P.S.F. Bellgowan, S. M. Rao, and R. W. Cox. 1999. Language processing is strongly left-lateralized in both sexes: Evidence from functional MRI. *Brain* 122: 199–208.

Geschwind, N., and A. M. Galaburda. 1985. Cerebral lateralization: Biological mechanisms, associations, and pathology: I. A hypothesis and a program for research. *Archives of Neurology* 42: 428–59.

Gur, R. C., B. I. Turetsky, M. Matsui, M. Yan, W. Bilker, P. Hughett, and R. E. Gur. 1999. Sex differences in brain gray and white matter in healthy young adults: Correlations with cognitive performance. *Journal of Neuroscience* 19: 4065–72.

Im, K., J-M. Lee, O. Yttelton, S. H. Kim, A. C. Evans, and S. I. Kim. 2008. Brain size and cortical structure in the adult brain. *Cerebral Cortex* 18: 2181–91.

Ivry, R. B., and L. C. Robertson. 1998. *The Two Sides of Perception*. Cambridge, MA: MIT Press.

Jaeger, J. J., A. H. Lockwood, R. D. van Valin Jr., D. L. Kemmerer, B. W. Murphy, and D. S. Wack. 1998. Sex differences in brain regions activated by grammatical and reading tasks. *NeuroReport* 9: 2803–07.

Ladavas, E., C. Umilta, and P. E. Ricci-Bitti. 1980. Evidence for sex differences in right-hemisphere dominance for emotions. *Neuropsychologia* 18: 361–66.

Mead, L. A., and E. Hampson. 1995. A selective effect of ovarian hormones on left visual field performance in verbal and nonverbal tachistoscopic tasks. *Journal of the International Neuropsychological Society* 1: 176.

Miller, R. 1996. *Axonal Conduction Time and Human Cerebral Laterality: A Psychobiological Theory*. Amsterdam: Harwood Academic.

Moes, P. E., S. B. Warren, and M. T. Minnema. 2007. Individual differences in interhemispheric transfer time (IHTT) as measured by event related potentials. *Neuropsychologia* 45: 2626–30.

Molfese, D. L. 1990. Auditory evoked responses recorded from 16-month-old human infants to words they did and did not know. *Brain and Language* 38: 345–63.

Natale, M., R. E. Gur, and R. C. Gur. 1983. Hemispheric asymmetries in processing emotional expressions. *Neuropsychologia* 21: 555–65.

Neblette, C. B. 1938. *Photography: Its Principles and Practice*. 3rd ed. New York: Van Nostrand, 45–46.

Okada, F., Y. Tokumitsu, Y. Hoshi, and M. Tamura. 1993. Gender- and handedness-related differences of forebrain oxygenation and hemodynamics. *Brain Research* 601: 337–42.

Pugh, K. R., B. A. Shaywitz, S. E. Shaywitz, R. T. Constable, P. Skudlarski, R. K. Fulbright, R. A. Bronen, D. P. Shankweiler, L. Katz, J. M. Fletcher, and J. C. Gore. 1996. Cerebral organization of component processes in reading. *Brain* 119: 1221–38.

Reite, M., J. Sheeder, P. Teale, D. Richardson, M. Adams, and J. Simon. 1995. MEG based brain laterality: Sex differences in normal adults. *Neuropsychologia* 33: 1607–16.

Rode, C., M. Wagner, and O. Güntürkün. 1995. Menstrual cycle affects functional cerebral asymmetries. *Neuropsychologia* 33: 855–65.

Sanders, G., and D. Wenmoth. 1998. Verbal and music dichotic listening tasks reveal variations in functional cerebral asymmetry across the menstrual cycle that are phase and task dependent. *Neuropsychologia* 36: 869–74.

Shaywitz, B.A., S. E. Shaywitz, K. R. Pugh, R. T. Constable, P. Skudlarski, R. K. Fulbright, R. A. Bronen, J. M. Fletcher, D. P. Shankweiller, L. Katz, and J. C. Gore. 1995. Sex differences in the functional organization of the brain for language. *Nature* (London) 373: 607–09.

Skrandies, W., P. Reik, and C. H. Kunze. 1999. Topography of evoked brain activity during mental arithmetic and language tasks: Sex differences. *Neuropsychologia* 37: 421–30.

Weekes, N. Y., L. Capetillo-Cunliffe, J. Rayman, M. Iacoboni, and E. Zaidel. 1999. Individual differences in the hemispheric specialization of dual route variables. *Brain and Language* 67: 110–33.

chapter 7

Neurology, psychiatry, and the female brain

The problem of mind-body dualism is a wonderful playground for philosophers. Because the existence, or nonexistence, of a soul cannot be "proven," the exercise becomes one of logical argument. In most circumstances, the problem can remain in the realm of the philosophers. Occasionally, however, the question of mind-body dualism intrudes upon the realm of real life, sometimes with tragic consequences. A prime example of this intrusion is in the assumptions often made regarding the disciplines of neurology and psychiatry.

The domain of neurology is brain pathology, the study and treatment of brain diseases and injury. The perceived domain of psychiatry, on the other hand, is sometimes a bit unclear. It includes the study and treatment of a disrupted or abnormal "mind" or "thought." Mind or thought . . . something removed from or beyond the physical substance of the brain? Here the majority of biologically oriented psychiatrists would say, "Rubbish, it is brain pathology. Disorders with a biological basis, amenable to biologically based treatments." Unfortunately, however, there are others (including some health-care professionals) who cling to the "disordered mind" perspective, apparently independent of biology, and once one removes the biology, then the ghosts and witches and prejudices are allowed to creep in. People who would never dream of saying to an insulin-dependent diabetic, "Get a grip, you're just using insulin as a crutch," will quite seriously make that remark to a depressed person who requires antidepressant drugs.

This chapter is divided into two main sections, neurological disorders and psychiatric disorders. The distinction is one of convenience and convention, not biology.

Neurology

It has been understood for a number of years that there are differences between females and males in the prevalence of some neurological disorders. For example, migraine headaches occur more frequently in females than in males, as does multiple sclerosis (MS). Amyotrophic lateral sclerosis (ALS) occurs more frequently in males than in premenopausal females

(4:1); however, the rate of occurrence is equal for males and postmeno-pausal females (Bromberg 1999). It has also been demonstrated that for some disorders the severity of the disease state changes with the menstrual cycle.

Chronic headaches

Headaches are experienced by more than 80 percent of premenopausal females, and it has been estimated that approximately 14 percent of females experience "significant," that is, very painful and temporarily debilitating, headaches more than four times per month. The International Headache Society (IHS) classification system categorizes headaches on a scale from one to twelve (Headache Classification Committee 1988). The classifications are divided into "primary" headaches (IHS Groups 1 to 4) in which the headache itself is the disorder, and "secondary headaches" (IHS Groups 5 to 12), which includes headaches that are related to another disorder such as trauma, infection, or substance withdrawal. IHS Group 1 is migraine headaches, which are suffered by approximately 25 percent of females and 8 percent of males. Migraine is divided into two types, those where an aura (e.g., flashing light patterns in the visual field, blind spots, temporary numbness) is experienced before the onset of the pain and migraine without aura. There is substantial evidence that estrogen, or more accurately, withdrawal of estrogen, is a migraine trigger in many females, with 33 percent of migraines being classified as "menstruation-related migraine." The majority of people with migraines experience the headaches for the first time after puberty, with the difference in headache frequency between females and males only becoming evident in the post-puberty years. During pregnancy the frequency and severity of migraines improves in the majority of females, approximately 58 percent; however, in a small number of females (around 2 percent) migraines occur for the first time in pregnancy (Ertresvag et al. 2005). Of females who experience nonmigraine headaches, around 25 percent report that the headaches disappear during pregnancy.

Another category of headache with a higher incidence in females is the "medication overuse" headache, IHS Group 8. This category includes headaches caused by drugs of abuse (including the classic "hangover") but also includes headaches caused by overuse of the medications used to treat headache. Medication overuse headache occurs in only 1 percent to 2 percent of the population, but the majority of individuals are female (Rossi et al. 2008). A recent study by Rossi and colleagues (2008) examined the relationship between endocannabinoids and serotonin in chronic migraine (n = 20), medication overuse headache (n = 20), and age-matched non-headache controls (n = 20). For this study, blood samples were collected from each of the participants and the platelets (the cells important for blood coagulation) were isolated from the samples for biochemical analysis. The samples were analyzed for two endogenous cannabinoids, anandamide and 2-acylglycerol, and serotonin. The results of the study showed that the cannabinoids and serotonin were significantly lower in the two headache groups compared with the control participants. The authors suggested that their results may explain females' reduced threshold for pain and also the increased female prevalence of migraine.

Stroke

A stroke occurs when an area of the brain is suddenly deprived of oxygen, either as a result of a blockage of a blood vessel (ischemic stroke) or as a result of a ruptured blood vessel (hemorrhagic stroke). The incidence of stroke is usually reported to be the same for females and males, although it has been suggested that there are sex differences in factors such as severity, symptoms, and extent of disability. A study of 505 patients presenting with their first ischemic stroke looked for female-male differences in the characteristics of their strokes (Barrett et al. 2007). There were 276 patients in the study, 45 percent females and 55 percent males ranging in age from 19 to 94 years. There were no differences between females and males in the amount of damage produced by the stroke, language deficits, visual disturbances, or numbness. Females were, however, significantly more affected by weakness than males. The range of ages included in this study is interesting because although the occurrence of strokes is generally associated with advancing age, strokes do occur in younger adults and even in children. Migraine with aura is reported to be risk factor for stroke in young females (Bousser 2004), although the incidence is quite low. The stroke apparently occurs during the aura, when constriction of the blood vessels associated with the migraine becomes so severe that the blood flow to part of the brain is interrupted. There may, however, be other interpretations of the "migrainous strokes." First, it is possible that the occurrence of an ischemic stroke may cause the aura followed by the migraine. Second, another disorder, such as a vascular disease, could be responsible for both the stroke and the migraine. A number of studies, using CT imaging or MRI, have reported a significant increase in abnormalities in the white matter of migraine sufferers compared with controls (see Barrett et al. 2007 for a review), which is an indicator that strokes have occurred.

Chronic pain

There is a good deal of evidence to suggest that the pain threshold is lower in females than in males, possibly by as much as 50 percent. There is also evidence to suggest that the pain threshold varies across the menstrual cycle, usually by about 5 percent to 10 percent (Rollman 1993). Unfortunately, it is very difficult to compare results across pain studies because of methodological differences, including such factors as sex stereotypes ("males endure pain no matter what") and demand characteristics (if one knows it is a "pain" experiment, one has certain expectations). In addition to human studies, however, there is evidence from experimental animal studies to support the reported sex differences in pain perception. Two studies by Coyle et al. (1995, 1996) have used partial sciatic nerve ligation in rats to produce tactile allodynia, a pain syndrome in which innocuous stimuli (such as a light touch with a cotton bud) are experienced as painful. In these studies, the variable measured was paw withdrawal from the stimulus. The authors report that when the responses of female rats and male rats were compared, the procedure produced abnormal paw withdrawal in 64 percent of female rats, but only 29 percent of male rats. In a further study, the responses of intact and OVX female rats were compared. The number of withdrawal responses of the intact rats was significantly greater than the withdrawal responses of the ovarectomized rats. Interestingly, this type of pain has also been reported to be more prevalent in human females than in human males.

A review of the literature on pain perception and the menstrual cycle analyzed the results of 16 studies on experimental pain in healthy humans. Different measures and methods were used in the studies, but the authors reported remarkably consistent results. For all types of experimental pain, except pain produced by electrical stimulation, the threshold was highest in the follicular phase. For pain induced by electrical stimulation, the threshold was highest in the luteal phase (Riley et al. 1999). There is evidence to suggest that the use of oral contraceptives may influence the perception of pain. Dao et al. (1998) have reported that in a study of females with myofascial pain, there was no difference in the pain experience across the menstrual cycle. However, the experience of pain was more severe and more constant in oral contraceptive users than in the normally cycling females.

Stress

Excessive stress is one of the most pervasive problems reported by adults in the Western world. Stress is, rightly or wrongly, blamed for all kinds of unwanted occurrences, for example, divorce, headaches, weight gain, weight loss, dry skin, limp hair . . . the list goes on. In some cases, the word "stress" seems to be used, somewhat trivially, to indicate our reaction to something that hasn't worked out as we expected. Unfortunately, this use of the word trivializes a process that is absolutely essential for survival. The term "stress" really refers to a number of physiological processes, triggered by internal or external events (stressors), that are crucial for saving life and preventing injury under threatening circumstances. This life-saving reaction is called the "fight-or-flight response" and it is a wonderful example of hormones in action. Imagine early woman gathering berries in a forest. The foliage is thick and tangled and she must reach far into the brambles to pick the fruit. She is aware that her situation is vulnerable and all her senses are on high alert. From nearby, to her left, she hears a rustling sound and the snapping of small branches. Her eyes involuntarily move in the direction of the sound as she sniffs the breeze coming from the same direction. Her visual system cannot deliver any useful information, but she smells the distinct scent of bear. Before she can consciously analyze the meaning of this information, she has dropped the berries she was holding and is running as fast as she can in the opposite direction. When she reaches safety she is out of breath, her heart is pounding, and she is shaking. She is also alive and uninjured. Her fight-or-flight response has served her well.

The hormonal system responsible for the fight-or-flight response is known as the Hypothalamic-Pituitary-Adrenal (HPA) axis. It includes the hypothalamus, the anterior pituitary gland, and the adrenal glands. The hypothalamus receives all kinds of information from the rest of the brain, including alerting signals from the reticular formation and the sensory systems. When a danger signal arrives in the hippocampus, it triggers the release of corticotrophin releasing hormone, a hormone that jolts the anterior pituitary gland into action. The anterior pituitary releases adrenocorticotrophic hormone (ACTH). The ACTH travels rapidly through the bloodstream and alerts the adrenal glands on top of the kidneys that flood the system with adrenaline and release corticosteroids. This cascade of hormones activates the sympathetic nervous system, causing a number of simultaneous actions, all geared to sustain a fight or a flight. The heart rate increases, sending a surge of blood to the brain and the large muscles. At the same time the blood flow to the extremities is reduced; if you get injured you will bleed less. Corticosteroids are released into the bloodstream, another practical measure; if you injure a limb the anti-inflammatory action of the steroids may reduce pain and swelling enough for you to escape. Blood sugar increases (extra energy), while gastrointestinal motility decreases (this is not the time to spend energy digesting a meal). Under normal conditions, the fight-or-flight response only lasts as long as it is needed, then the parasympathetic nervous system is activated and homeostasis is restored. As an emergency measure, this cascade of hormones does no harm. If, however, the response is prolonged then adverse responses can occur including damage to the cells in the hippocampus. This is "stress." Many factors may cause prolonged stress: the death of a friend, family member, or beloved pet; divorce (or marriage); a change or loss of employment; illness; war . . . the list goes on and on. The ability to handle stress is, ultimately, an important survival factor.

It can be difficult to measure the behavioral effects of stress. Usually a person undergoing a stressful situation is not in a position to sit down and complete a stress questionnaire or a series of cognitive tasks. For this reason, laboratory studies of stress-induced behavior have to employ an ethical but stressful stimulus. In rats, restraint is an excellent stressor. For humans, the experimenter has to be a bit more resourceful. One method of inducing stress is to ask the subject to relate in intricate detail a remembered highly stressful situation and to describe her feelings at the time of the event, and her feelings on recounting the event. Interestingly, an early study measuring stress using this method did not find any changes in the stress hormone cortisol related to the phases of the menstrual cycle (Abplanalp et al. 1977). A recent study used a very different approach, asking subjects to rate the intensity of faces displaying fear and disgust — that is, stress — in someone else. Fifty-two subjects were tested early in the cycle when progesterone was low and again later in the cycle when progesterone was high. With high levels of progesterone, the subjects rated the two test faces as more intense than control faces, and as more intense than when progesterone was low. The authors interpret this result as indicative of a protective response as raised progesterone can signal pregnancy (Conway et al. 2007). Another source of stress is drug withdrawal, as used in a study of stress responses in 12 cocaine-dependent females during the first month of withdrawal, compared to 10 nondependent females (Fox et al. 2008). For this study daily samples of saliva were collected and analyzed for cortisol, estrogen, and progesterone. The cocaine-dependent females had consistently higher levels of cortisol and progesterone than the controls across the entire cycle and they also, as expected, reported a much more negative mood. Cocaine craving did not alter across the cycle.

To get firm empirical data on sex differences in response to stress, some of the best work comes from experiments with laboratory rats. Here there is consensus: numerous studies have demonstrated that female rats are more resistant to stress-induced cognitive impairment — for example, memory impairment on a spatial memory task — than male rats (e.g., Bowman 2005).

Stress during pregnancy has been reported to be linked with mixed-handedness in the child (Glover et al. 2004). A study of 7,431 mother-child pairs measured anxiety in the mother at 18 and 32 weeks gestation and the handedness of the child at 42 months. Analysis of their results revealed a significant relationship between maternal anxiety at 18 weeks of pregnancy with mixed-handedness in the child (the majority of the mixed-handed children were male). A later study of one hundred and ten 6-year-old children and their mothers failed to confirm a stress effect in early pregnancy but did report that stress in late pregnancy (weeks 37–38) was associated with mixed-handedness (Gutterling et al. 2007). A study of 49 adolescents (age 17) reported that the subjects in the study whose mothers who had experienced stress during weeks 12 to 22 of pregnancy performed less well in experiments requiring integration of complex tasks and materials than the control group whose mothers were not stressed (Mennes et al. 2006). This result, the authors suggested, is related to subtle development disturbances in the orbitofrontal cortex.

Sleep

Sleep disturbances are the bane of many people, female and male. They may be associated with anxiety or depression, or they may occur on their own. There are different types of disturbances: some people may have trouble falling asleep but sleep well when they eventually do go to sleep; others may fall asleep easily then wake in the early hours and stay awake until just before the alarm goes off, then fall into a deep, peaceful sleep. There is yet another group for whom sleep is patchy throughout the night, waking every hour or so then falling asleep again. Whatever the form of sleep disturbance, the outcome is usually chronic tiredness with daytime sleepiness and a craving for naps. Females consistently report more sleep problems than males. In the past this has been, at least partially, attributed the females' role in childcare and household responsibilities, including in rural communities rising before everyone else to prepare breakfast for family and farmhands, and to start food preparation for the rest of the day. Until fairly recently, sleep studies were conducted almost exclusively in males, so the female perspective wasn't even represented (see Dzaja et al. 2005 for a review). Normal sleep is composed of rapid eye movement (REM) and non-REM sleep, during which a period of slow wave sleep occurs. REM sleep is characterized by cortical activity similar to that observed during waking. During non-REM sleep everything slows down and energy consumption is at its lowest. It has been known for some time that slow wave sleep time decreases with increasing age, particularly in males, while elderly females have fairly well-preserved slow wave sleep. This difference starts to develop after about age 30. Although some studies show small cycle-related effects, the majority of evidence suggests that in healthy females, there are few menstrual cycle–related changes in sleep. In addition, when cycle-related sleep disturbances are reported, they are usually associated with other cycle-related disturbances such as premenstrual dysphoric disorder or depression. During pregnancy total sleep time increases during the first trimester then decreases toward the female's usual sleep time by the third trimester (Hedmann et al. 2001). Sleep is consistently reported to be disturbed during menopause, with the primary reason for the disturbance being hot flushes. After menopause total sleep time tends to decrease with the normal time for going to sleep and for waking shifting to approximately an hour earlier (see the section on Alzheimer's disease and aging).

Multiple sclerosis

MS is often referred to as "the disease of young adults." It seldom occurs before puberty or after the age of 50. Across this 35-year age span the female-male ratio remains the same (approximately 2:1). There is a genetic component, possibly as a predisposing factor, to the development of MS. A study looking at concordance rates for parent-child pairs has reported that out of 75 pairs concordant for MS, the pairings were distributed as follows: mother–daughter (n = 40); father–daughter (n = 21); mother–son (n = 13); father–son (n = 1) (Sadovnick et al. 1991). The cause of MS has not been determined. Even after controlling for the normal female/male disease rate, there is still a female bias to the inheritance (Figure 7.1). MS is a demyelinating disease, generally held to be of autoimmune origin. Some yet-to-be-determined factor causes the activation of T-lymphocytes sensitized to myelin basic protein. The activation of these T-cells (Th1 cells) results in a cascade of cytokines, the chemical messengers involved in immune/inflammatory responses. At the same time, another group of T-cells (Th2 cells) that produce anti-inflammatory cytokines are also activated. Under normal conditions, the Th1 cells perform the necessary immune response — for example, recognizing and destroying bacterial

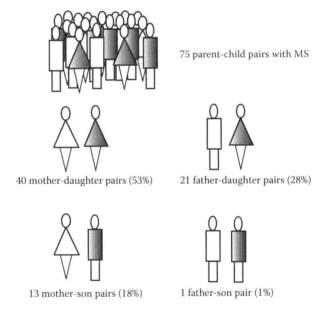

75 parent-child pairs with MS

40 mother-daughter pairs (53%)   21 father-daughter pairs (28%)

13 mother-son pairs (18%)   1 father-son pair (1%)

Figure 7.1 Concordance for multiple sclerosis in parent–child pairs. Data from Sadovnick et al. (1991).

invaders — then Th2 cells stop the inflammatory response and initiate the
healing process. It is a finely tuned system of checks and balances with
multiple feedback mechanisms (Figure 7.2).

Once they have been activated, lymphocytes cross the blood brain
barrier and ultimately attack the myelin sheaths surrounding the axons.
In the CNS myelin is composed of the processes of a particular type of
glial cell, oligodendrocytes. In the early stages of the disease, only the

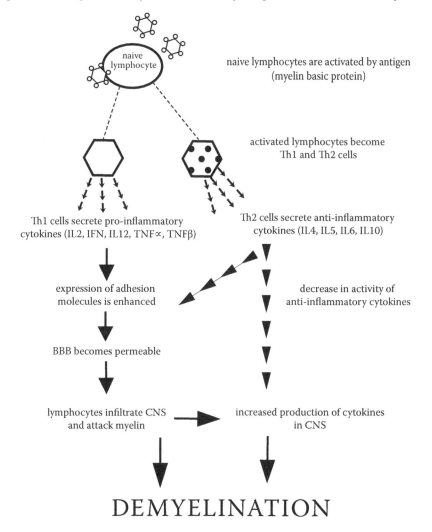

Figure 7.2 The cytokine cascade leading to demyelination.

cell processes producing the myelin sheath are damaged. The cell body remains intact and able to repair the myelin. Later in the disease, however, the oligodendrocytes begin to die, and at that stage, the lesions become permanent. Demyelination results in a disruption of neural transmission, and the conduction of action potentials along the axons is slowed. The symptoms produced by this slowing of action potentials will depend upon which axons are demyelinated at any particular time. It is this constant state of flux that characterizes MS. Symptoms come and go. One symptom, for example, tingling in the fingers of one hand, will disappear only to be replaced by another — for example, weakness in one leg. A diagnosis of MS requires that the symptoms be "disseminated in time and space." There are two types of MS: relapsing-remitting (RR-MS) and progressive (P-MS). RR-MS is characterized by the appearing and disappearing symptoms as described above.

Experimental allergic encephalomyelitis (EAE) is an animal model of MS that resembles the human disease. It is produced in rats or mice by raising an immune response to myelin basic protein (MBP). An early study looked at the effects of hormones on EAE. Arnason and Richman (1969) administered oral contraceptives to EAE female rats. The authors found that although 8/10 of the control rats developed EAE within three weeks of inoculation, only 2/10 rats treated with the oral contraceptive Enovid (estradiol) and 3/9 rats treated with the oral contraceptive Provest (estradiol plus progesterone) developed EAE. When the estradiol and progesterone constituents of Provest were administered separately to additional groups of rats, it was found that only the estradiol inhibited the development of EAE. The precise form that EAE takes differs between the species used, and so, depending upon the question being asked, the choice of species will be crucial. In addition, the effects of hormones show species differences. EAE in the SJL mouse closely resembles human MS, including greater disease susceptibility in females and a relapsing-remitting disease course. Voskuhl et al. (1996) used a preparation of myelin basic protein from guinea pig spinal cord to induce MBP activated T-cells in both female and male SJL mice (the donors). Following a two-week incubation period, the lymph nodes were removed from the inoculated mice. The cells were removed from the lymph nodes, washed, and incubated in a MBP culture for four days. Finally, the activated T-lymphocytes from the female donors were injected into one group of female and one group of male mice. T-lymphocytes from the male donors were injected into two other groups of female and male mice. The treated mice developed EAE. The female mice developed the disease faster and with greater severity than the male mice. A suspension containing 5×10^7 activated T-cells produced maximum disease severity in females, but did not produce the severe disease state in the male mice. The effect was the same whether the injected cells came from female or male donor mice. This result suggests that, at least in SJL mice with EAE, the females were more vulnerable to autoimmune attack than the males.

At the cellular level, it has been reported that when human helper T-cells are incubated in the presence of progesterone, the cells produce significantly greater amounts of Th2 (anti-inflammatory) cytokines, compared to the same cell line incubated under the same conditions, but without the addition of progesterone (Piccinni et al. 1995). This suggests that the presence of progesterone should be beneficial in MS patients. Another study, this time comparing T-cells from people with MS with T-cells from control subjects, has reported that when the cells were incubated with estradiol, the secretion of IFN-γ (inflammatory cytokine) was increased (Gilmore et al. 1997). The secretion of TNF-α (inflammatory cytokine) showed a concentration-dependent modulation, an increase at low concentrations but a decrease at high concentrations. There was no effect on the secretion of IL-4 (anti-inflammatory cytokine), but, just to make things interesting, the secretion of IL-10 was increased. There was no difference between the cells obtained from normal donors and from MS patients. From this study it appears that estradiol can have both inflammatory and anti-inflammatory effects, making predictions very difficult.

At the clinical level, fluctuating hormone levels have been well demonstrated to play a role in MS. For example, a survey of 72 females with MS has reported that of the 60 patients with RR-MS, 23/45 (51 percent) normally menstruating patients reported the symptoms to be worse during the premenstrual or early menstrual phase; 22/45 (48 percent) reported no cycle-associated change (Zorgdrager and de Keyser 1997). Of the patients taking oral contraceptives, 3/15 (20 percent) with RR-MS reported that their symptoms worsened premenstrually/menstrually, while 12/15 (80 percent) reported no change. Twelve patients in the study had P-MS, and none of these patients reported cycle-associated changes in their disease state (Figure 7.3).

Two studies have used MRI to examine changes in MS lesions associated with different phases of the menstrual cycle. Interestingly, the two studies reported apparently conflicting results. Pozzilli et al. (1999) used gadolinium-enhanced MRI to measure the disease activity in eight female patients (mean age 34 years) with RR-MS (mean disease duration seven years). A monthly MRI scan was performed on each patient for four consecutive menstrual cycles. For half of the patients the first scan was performed in the follicular phase (days 3–9), with the following scans performed in the luteal phase (days 2–28), follicular phase, and luteal phase. For the remaining patients, the pattern of scans began with the first scan in the luteal phase then alternated for the remaining scans. This pattern of scanning resulted in two follicular phase scans approximately two months apart and two luteal phase scans also approximately two months apart. On the day of the first scan blood was collected and stored at −20° C; the

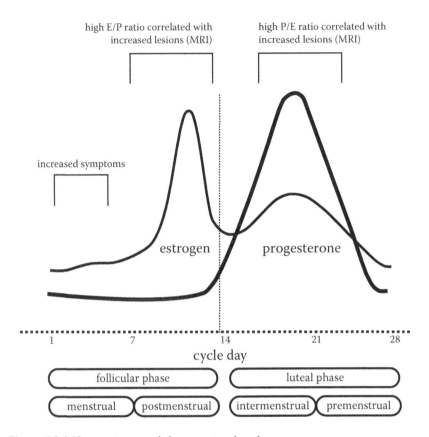

Figure 7.3 MS symptoms and the menstrual cycle.

blood samples were later analyzed for follicle stimulating hormone, leutinizing hormone, 17-beta-estradiol, and progesterone. On analysis, progesterone was detectable in the luteal phase only. MRI analysis showed that there was no difference in the frequency or volume of lesions between the follicular and luteal phase. There was, however, a significant positive correlation between the ratio of progesterone to estradiol and both the number and volume of observed lesions in the luteal phase (Figure 7.3).

Bansil et al. (1999) also used gadolinium-enhanced MRI to study menstrual cycle–associated changes in MS patients. Thirty MS patients participated in the study (20 RR-MS, 10 P-MS). Each patient received one MRI scan, and a blood sample was also taken on the day of the scan. The patients were divided into three groups based upon their serum hormone levels at the time of testing: Group 1, low estradiol, low progesterone (early follicular, n = 14; RR-MS = 10, P-MS = 4); Group 2, high estradiol, low progesterone (late follicular, n = 6; RR-MS = 2, P-MS = 4); Group 3, high progesterone but variable estradiol (luteal, n = 10; RR-MS = 6, P-MS = 4). It is interesting that the patients were recruited for testing based on the day of the menstrual cycle, apparently to provide equal group sizes. Analysis of blood hormones, however, revealed that for some patients the hormone levels were not consistent with the reported cycle day. As a result, all of the subjects were grouped by hormone levels on the day of testing. Analysis of the MRI scans revealed that there was a significantly higher number of lesions for Group 2, high estrogen/low progesterone, compared with Group 1. In addition, there were significantly more lesions with a high estrogen to progesterone ratio (Group 2) compared with a low estrogen to progesterone ratio (Group 1 plus Group 3) (Figure 7.3). In considering how these results relate to the results of Pozzilli et al., a potentially confounding factor needs to be considered. The Pozzilli et al. study included only patients with RR-MS; Bansil et al. included patients with P-MS. In Group 2 (the smallest sample), there were only two RR-MS patients but four P-MS patients. No one knows if there are hormonal differences between people with RR-MS and P-MS; however, Zordrager and Keyser (1997) reported that P-MS was not associated with cycle-related changes. Because the profile of P-MS is substantially different from RR-MS, the possibility of an experimental confound must be considered.

It has been reported frequently that during pregnancy, when progesterone is continuously present, there are fewer relapses, and disability scores tend to be lower. Birk et al. (1990) followed the pregnancies of eight patients with MS and continued their observations until six months postpartum. Six healthy pregnant females, who did not suffer from MS, were also observed for the same period. None of the patients experienced relapses during pregnancy. The number of CD8 suppressor T-cells (a measure of immune system activity) was lower during pregnancy and the ratio of CD4 helper to CD8 suppressor T-cells was higher, indicating reduced disease activity. Within seven weeks of delivery, 6/8 MS patients had experienced at least one relapse. This is a rate of relapse of 75 percent in the postpartum period. A similar result in a study of 269 MS pregnancies has been reported by Confavreux et al. (1998). For the patients in this study, the mean relapse rate in the year before pregnancy was 0.7 ± 0.9 relapses per patient. By the third trimester the rate of relapse had dropped to 0.2 ± 1.0 relapses per patient. In the first three months after delivery, the relapse rate rose significantly to 1.2 ± 2.0 before returning to the prepregnancy baseline. An MRI study of two MS patients, one scanned monthly and the other at three-month intervals, demonstrated a reduction in active lesions, which correlated with the decreased clinical signs of the disease during the second half of the pregnancy (van Walderveen et al. 1994). Finally, it has been reported in a survey of 19 postmenopausal MS patients, that 54 percent experienced a worsening of their symptoms after menopause while 8 percent reported that their symptoms had diminished (Smith and Studd 1992). These same patients reported that their condition

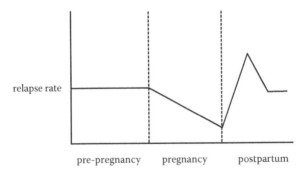

Figure 7.4 MS symptom relapse rate associated with pregnancy and childbirth.

either improved (75 percent) or did not change (25 percent) with hormone replacement therapy. Also in this study, 9/11 premenopausal MS patients reported that their symptoms were worse premenstrually and 2/11 reported that their symptoms improved at that time (Figure 7.4).

The results of the experiments discussed in this section are summarized in Table 7.1. Examination of the table reveals nothing but two columns of apparently contradictory results, at least in terms of the absolute values of estrogen and progesterone. After close study of Figure 7.3, a possible pattern begins to emerge, however. Perhaps it is not only the absolute values of the two hormones that are important, but also the changes in hormone levels, particularly estrogen levels. During the luteal phase and particularly premenstrually, the levels of both progesterone and estrogen are fluctuating rapidly. During the menstrual phase there is a period of stability, then estrogen begins to rise rapidly to its pre-ovulatory peak and then plummet a few days later. Pozzilli et al. tested their patients between days 3 to 9 (follicular) and days 21 to 28. If the patients were tested earlier in the follicular phase (the estrogen level does not begin to rise until day 8), when there had been five or six days of relative hormonal stability, the symptoms might be expected to be less severe than on days 21 to 28 when both progesterone and estrogen levels were falling. Bansil et al. tested their subjects on days 1–3 (Group 1), 14–16 (Group 2), and 21–23 (Group 3), and then adjusted their groups according to serum hormone levels. Between days 14 to 16, just after the estrogen peak, when they reported the greatest number of lesions, estrogen levels dropped rapidly. Between days 21 to 23, progesterone was dropping sharply and estrogen was decreasing slowly. Relatively speaking, Group 2 was in the most unstable period.

Interestingly, MS symptoms usually become better or disappear entirely during pregnancy when both estrogen and progesterone are high. This pregnancy-related change has been recognized for many years, and there are a number of anecdotal reports in which doctors treating MS

Table 7.1 Summary of estrogen and progesterone effects on MS.

Variable	Estrogen	Progesterone
EAE in rats	Beneficial	No effect
T-cells in vitro	Mixed	Beneficial
Pregnancy		Beneficial
Menstrual cycle studies		
MRI/RR (8) Pozzilli et al.		Detrimental
MRI/RR (20) & P (10) Bansil et al.	Detrimental	

	Day 1–5	Day 6–14	Day 15–21	Day 22–28
Self-report/RR (23/45)	Worse at start			Worse 26–28
Self-report/P (12/12)	No change	No change	No change	No change
Self-report (9/11)				Worse
Self-report (2/11)				Better

patients have suggested that pregnancy might be a good way to obtain relief, albeit temporarily, from the symptoms.

Epilepsy

Epilepsy is not a single disorder but is, in fact, a cluster of disorders characterized by abnormal and excessive (epileptiform) activity in the brain, consisting of synchronized burst firing of populations of neurons. While the defining feature of epilepsy is this epileptiform activity, as diagnosed using EEG, the clinical signs of a seizure can range from the gentle twitching of a single finger (simple partial seizures) to full-blown generalized seizures with loss of consciousness (generalized tonic-clonic seizures). The incidence of epilepsy is usually reported to be the same for females and males. However, sex differences in the expression of epilepsy have long been recognized and the hormonal basis of these differences is also understood. It is estimated that between 12 percent and 78 percent (depending upon the criteria applied) of females with epilepsy experience changes in the severity or frequency of seizures associated with the menstrual cycle. There is even a name for seizures that occur primarily or with increased frequency during a particular phase of the menstrual cycle: catamenial epilepsy (Zahn 1999).

Although epileptiform activity can occur in any area of the brain, one of the regions most often affected is the hippocampus. Temporal lobe epilepsy (TLE, partial complex seizures) is a common form of the disease, and in TLE the locus of the epileptiform activity is often in the hippocampus. It has been demonstrated that a distinctive pattern of cell loss in the hippocampus (mesiotemporal sclerosis) is associated with TLE (Woolley and Schwartzkroin 1998). Estradiol has been demonstrated to induce synaptic plasticity in the hippocampus by promoting the formation of new synapses. Significantly, the receptors in these newly induced synapses are primarily the N-methyl-D-aspartate (NMDA) subtype of the glutamate receptor, the receptor subtype whose overactivation is associated with hyperexcitability and cell death.

The kindling model of epilepsy uses electrical or chemical stimulation of an area of the limbic system, for example, the amygdala, to produce epileptiform activity in an animal (e.g., rat). To deliver the stimulation, electrodes (or a cannula, in the case of chemicals) need to be positioned in the target brain region, on one or both sides of the brain, depending upon the experimental question. In addition to delivering the stimulus, electrodes are often used to record the brain activity resulting from the stimulation. In the early stages of the kindling, the stimulation will produce a short burst of neuronal activity (the after-discharge [AD]). With repeated stimulation the AD becomes prolonged and may spread to neighboring brain regions. In the most extreme case, the epileptiform activity may spread to the limbic structures on the opposite side of the brain; hence the term "kindling" (Figure 7.5). Ultimately the rats exhibit generalized tonic-clonic seizures. Treatment of rats with estradiol has been demonstrated to potentiate the kindling process by reducing the number of stimulations required to produce generalized seizures (Wooley and Schwartzkroin 1998). In addition, it has been demonstrated that when estradiol binds to

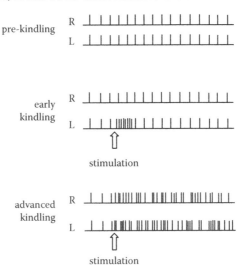

Figure 7.5 Schematic representation of kindling in the limbic system. Each vertical line represents an action potential of an amygdala neuron.

Table 7.2 Actions of estrogen and progesterone in animal models of epilepsy.

Estrogen	Progesterone
Lowers ECS threshold	Raises ECS threshold
Prolongs duration of seizures	Reduces seizure activity (including alcohol withdrawal)
Increased severity of seizures	Raises seizure threshold
Creates epileptiform foci with direct application	Causes sedation and anesthesia (also in humans)

Source: From Morrell (1999). ECS, electroconvulsive shock.

the steroid-binding site on the GABA$_A$ receptor, it reduces the chloride influx that normally occurs when GABA binds to the receptor, resulting in a decrease in the inhibitory effect of GABA. When progesterone binds to the steroid-binding site on the GABA$_A$ receptor, there is an increase in the chloride influx associated with GABA binding and, therefore, an increase in the inhibitory actions of GABA (Morrell 1999). There is substantial evidence supporting a seizure-promoting activity for estrogen and a seizure-reducing activity for progesterone (Table 7.2).

Females with epilepsy consistently report that stress and the menstrual cycle, in that order, are the two most consistent precipitants of seizures (Spector et al. 2000). There is a good deal of empirical evidence to support their reports. There are three patterns of seizure activity associated with catamenial epilepsy. Seizure activity may be worse in the three days before and the three days after the onset of the menstrual phase; at this time estrogen is low but progesterone is virtually nonexistent. Seizure activity may be worse at the time of ovulation, when estrogen is high. In cycles where ovulation does not occur (anovulatory cycles), estrogen remains relatively high but progesterone does not rise and seizures may occur apparently randomly in the luteal phase (Figure 7.6). In normal females and females with epilepsy who experience generalized seizures, less than 10 percent of cycles are anovulatory; however, in females with epilepsy who experience temporal lobe seizures, it has been reported that the rate of anovulatory cycles is around 33 percent (Zahn 1999). Other sex-related differences in temporal lobe epilepsy have also been reported. A PET study of 48 people with epilepsy (21 males, 27 females) has reported that in females, between seizures, there was a decreased glucose utilization (hypometabolism) within the temporal lobe contralateral to the area where the seizures begin (Savic and Engel 1998). With seizure onset, the epileptiform activity rapidly spread to this contralateral area where hypometabolism was observed. In males, on the other hand, the hypometabolism observed between seizures was primarily in the ipsilateral frontal areas and epileptiform activity spread to this area with seizure onset.

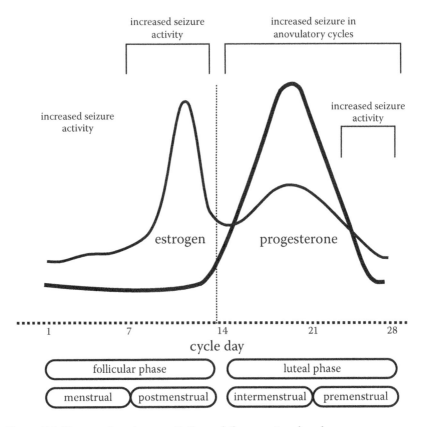

Figure 7.6 Changes in seizure activity and the menstrual cycle.

Pregnancy and epilepsy are a difficult, but not unmanageable, combination. In order to maintain a healthy pregnancy and to control her epilepsy, planning a pregnancy well in advance is essential for a female with epilepsy. Many neurologists who treat epilepsy recommend that planning for a pregnancy begin at least a year in advance. All the factors associated with pregnancy can be considered and the prospective mother can have her drugs adjusted and, if necessary, changed to provide optimal seizure control at the lowest possible drug dose. Most antiepileptic drugs have the potential to cause birth defects, although some are safer than others. For this reason, a change in drug management may be desirable and this, in itself, may require several months.

Parkinson's disease

Parkinson's disease (PD) was first described by James Parkinson in 1817 as the "shaking palsy." The "official" name given by the World Health

Organization is *paralysis agitans* (the Latin translation of "shaking palsy"). It is characterized by a resting tremor (4–6 Hz), bradykinesia (slowness of movement), and rigidity (the affected muscles appear to be constantly contracted). Cognitive deficits are also associated with PD in some patients. PD is associated with the loss of dopamine-containing neurons in the *pars compacta* region of the substantia nigra. The motor symptoms described above are a result of the loss of dopamine in the nigro-striatal pathway, which has profound effects on neural transmission in the basal ganglia (Figure 7.7).

The basal ganglia consist of four nuclei: the striatum, globus pallidus, substantia nigra, and subthalamic nucleus. The striatum is the control center for the basal ganglia (Parent et al. 2000). It is the major target for inputs to the basal ganglia from other parts of the brain. It projects to the output nuclei of the basal ganglia and these projections are entirely GABAergic. The GABA-releasing neurons in the striatum have dopamine receptors. The neurons projecting to the internal segment of the globus pallidus (GPi) and the reticular section of the substantia nigra (SNr) have type I dopamine receptors (D_1) (which, according to Parent et al., facilitate cell firing), while the neurons projecting to the external segment of the globus pallidus (GPe) contain type II dopamine receptors (D_2) (which, according to Parent et al., inhibit cell firing). Under normal conditions, the GABAergic cells in the striatum receive DA projections from the *pars compacta* of the substantia nigra (SNc). In Parkinson's disease it is the dopamine-containing neurons in the SNc that degenerate, reducing the dopamine input to the striatum. When the striatal neurons projecting to the GPi lose their inhibitory dopamine input, they increase their release of GABA and therefore their inhibition of the thalamus. This thalamic inhibition is thought to account for the impaired movement and the other classical signs of Parkinson's disease.

The age of onset of PD is usually over 60; however, there is a rare early onset form of the disease that may begin as early as 30 years of age. The prevalence of PD is estimated to be 1 percent of the population over 50 years of age. PD occurs more frequently in males than in females. The age of onset of the disease, the severity, and the time course are the same in the two sexes. Interestingly, females with PD have the same life expectancy as males with PD. In other words, since females normally have a greater life expectancy than males, PD decreases the life expectancy of females disproportionately (Diamond et al. 1990). Because PD is a disease normally associated with older age groups, the question of menstrual cycle effects on PD has seldom occurred. There are, however, a small number of reports that suggest that circulating hormones may affect disease severity. For example, in response to a questionnaire administered by mail to 352 females with PD, 75 percent of the patients not taking oral contraceptives reported a worsening of symptoms before and during menstruation; 48 percent of

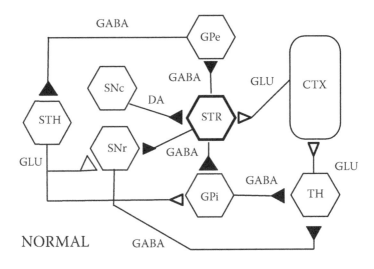

NORMAL

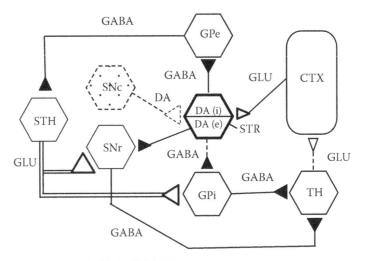

PARKINSON'S DISEASE

Figure 7.7 Schematic representation of changes in neurotransmitter activity in basal ganglia pathways associated with Parkinson's disease. Filled triangles represent inhibitory synapses; open triangles represent excitatory synapses. Broken lines indicate decreased activity in pathway. CTX, cortex; DA, dopamine; GLU, glutamate; GPe, globus pallidus external segment; GPi, globus pallidus internal segment; SNc, substantia nigra *pars compacta*; SNr, substantia nigra *pars reticulata*; STH, subthalamic nuclei; STR, striatum; TH, thalamus. DA (i) indicates inhibitory effect of DA on GPe projection neurons. DA (e), excitatory effect of DA on GPi and SNr projection neurons. Adapted and modified from Parent et al. (2000).

patients taking oral contraceptives reported the same menstrual phase–associated exacerbation in symptoms (Thulin et al. 1996). A study of 138 postmenopausal PD patients has reported a beneficial effect from estrogen replacement (Saunders-Pullman et al. 1999). The patients in this study had a disease duration of less than five years and had never taken L-Dopa (the most widely used treatment for Parkinson's disease). Thirty-four patients had received estrogen supplements (estrogen alone = 11; estrogen plus progesterone = 20; unknown formulation = 3). Analysis of the results indicated that there was a negative correlation between estrogen replacement and the score on the Unified Parkinson's Disease Rating Scale. A subset of people with PD develop dementia. A retrospective study of 87 females with PD but without dementia and 80 females with PD and with dementia reported a significantly greater incidence of dementia in patients not taking estrogen replacement (Marder et al. 1998), although the overall prevalence of PD was not affected by estrogen replacement.

Not surprisingly, there is a very small literature available on pregnancy in PD. Only about 3 percent of females with PD are still in their reproductive years. A number of studies have reported that PD symptoms worsen in cycle days 21 to 28 (see Rubin 2007 for a review). Interestingly, many of these same females reported that their menstrual cycle symptoms worsened after they developed PD. In a study of 17 pregnancies in PD, it has been reported that 11 of the females reported more rapid disease progress during pregnancy (Golbe 1987). It was reported in another study that the non-motor symptoms of PD seemed worse during pregnancy than the motor symptoms (Roy 2000, as cited in Rubin 2007).

Alzheimer's disease and aging

The normal aging process, which really begins about the time that development finishes, is accompanied by a mild, gradual decline in cognitive function very late in life. The changes include memory deficits, and a decline in abilities such as problem solving, visual-spatial orientation, and verbal fluency. There is a change in sleep patterns with an increase in stage 1 slow wave sleep and a decrease in rapid eye movement sleep. There is a tendency to sleep less and to wake more frequently, and many elderly people suffer from chronic sleep deprivation as a result of these changes. These behavioral changes are accompanied by a decrease in total brain weight, with some region-specific neuronal loss. There are decreases in the levels of dopamine, noradrenaline, and acetylcholine. On histological examination of aged brains, small numbers of senile plaques and neurofibrillary tangles can also be observed. At some stage, in some individuals, the normal aging process becomes a pathological process and dementia develops. There is recent evidence to suggest that variants in estrogen receptors are associated with an increased risk of non-dementia type cognitive impairment (Yaffe et al. 2007).

The most common form of dementia is Alzheimer's disease (AD). AD occurs more frequently in females than in males, with a ratio of approximately 2:1. It affects around 7 percent of the population over age 65, increasing to around 40 percent of the population over 80. There is a less common form of AD with early onset, between 30 and 50 years of age (see below). In the early stages of AD, it is difficult to distinguish the disease state from the normal aging process. However, as the disease progresses, there is a rapid decline in cognitive function until the patient becomes unresponsive and bedridden.

The neuropathology of AD is characterized by a loss of neurons resulting in an apparent "shrinkage" of the brain and enlargement of the ventricles. Cell loss is particularly pronounced in the hippocampus, amygdala, entorhinal cortex, anterior thalamus, locus coeruleus, raphe nuclei, and neocortex. On histological examination, the neuropathological changes that are the hallmarks of AD, namely cell loss, neurofibrillary tangles, and senile plaques, may be observed. The first signs of AD are usually observed in the entorhinal cortex; however, as the disease progresses, plaques and neurofibrillary tangles spread to the other vulnerable areas listed above and may ultimately be seen throughout the brain. The intracellular components of AD, neurofibrillary tangles, are composed of chains of *tau* protein that collect in the cells' cytoplasms. Tau is normally a soluble protein associated with normal cellular structure. In AD, however, the tau protein is much less soluble and collects in the intracellular space, disrupting cellular function and ultimately leading to cell death. Senile plaques are the extracellular component of AD pathology. They are deposits of *amyloid* that occur in the extracellular space and in the walls of blood vessels in the CNS. The main constituent of amyloid is β *amyloid,* and alterations in the expression of β amyloid have been demonstrated in both AD and Down's syndrome. As a result of the neuropathological changes, loss of acetylcholine-producing neurons in the nucleus basalis, the medial septum, and the diagonal band of Broca results in a widespread loss of acetylcholine. Other neurotransmitters, including dopamine and noradrenaline, are also decreased.

In approximately 10 percent of AD cases, the disease onset is before the age of 50 (early onset). Early-onset AD is considered to be familial (inherited). In addition, there are also genetic risk factors for late-onset AD. To date, five mutations have been identified that are associated with AD (Table 7.3). In approximately 30 percent of cases of early onset AD, there is a mutation of the gene encoding presenilin 1. Although the role of presenilin is not understood, it has been demonstrated that mutations affecting amyloid precursor protein (the protein from which β amyloid is derived) and presenilin result in the production of the damaging forms of β amyloid. Another protein that has also been associated with AD is

Table 7.3 Genetic mutations associated with AD.

Chromosome	Element Affected	AD Type
21	Amyloid precursor protein	Early onset/familial
14	Presenilin 1	Early onset/familial
1	Presenilin 2	Early onset/familial
19	Apolipoprotein E	Early onset/sporadic
		Late onset/familial
12	Alpha-2 macroglobulin	Early onset/sporadic
		Late onset/familial

apolipoprotein E (ApoE). The ApoE locus on chromosome 19 can express three different forms of ApoE (alleles), and one of these alleles, E4, is expressed in 50 percent of AD patients with the late-onset form of AD.

There is experimental evidence to support a protective role for estrogen in AD. For example, it has been demonstrated in vitro using a mouse hippocampal cell line, that estrogen can have antioxidant actions (Behl et al. 1995). Batches of cells were incubated with one of three different neurotoxins: amyloid β protein, hydrogen peroxide, or glutamate. For each neurotoxin, subgroups were pre-incubated with 17-β estradiol, progesterone, aldosterone, corticosterone, or cholesterol. The cells pre-incubated with 17-β estradiol were protected against the neurotoxins; none of the other hormones showed this protective effect. The neuroprotective action of 17-β estradiol was independent of estrogen receptors. Estrogen has also been reported to play a role in the metabolism of amyloid precursor protein, from which the material comprising senile plaques (β-amyloid protein) is derived (Jaffe et al. 1994). β-amyloid protein is an essential constituent of all cells; it is the abnormal accumulation in AD which is associated with cell death. Samples of a human-derived cell line that expresses estrogen receptors were incubated for 11 days in medium containing physiological levels of 17-β estradiol or in control medium. Following the incubation period, the medium in which the different batches of cells had been incubated was analyzed for the presence of a by-product of amyloid precursor protein metabolism. The levels of the by-product were significantly higher in the samples incubated with 17-β estradiol, suggesting that in these cells the metabolism of the protein had been enhanced.

One of the most difficult aspects of aging research is trying to obtain data in a retrospective study. No matter how cooperative the subjects and their caregivers may be, information becomes distorted (not to mention forgotten) with time. When the disease of interest is one characterized by cognitive deterioration and memory loss, as in AD, the problems become almost insurmountable. The Baltimore Longitudinal Study of Aging (BLSA) is a multidisciplinary study of the normal aging process that has followed

2000 individuals for almost 40 years. Every two years, participants in the BLSA undergo physical, neurological, and psychological assessment, including collection of information on the use of estrogen replacement (ER) or any other drugs or supplements. In addition to providing invaluable data on the normal changes associated with the aging process, BLSA also provides an excellent database for surveying the neuropathology of aging. For example, the data from 472 postmenopausal women in the BLSA who had already been followed for 16 years were analyzed for the incidence of AD (Kawas et al. 1997). Over the 16-year period, approximately 45 percent of the subjects used ER. Also over that period, 34 subjects developed AD, including 9 subjects who had taken ER. The data were analyzed to estimate the risk of developing AD when taking ER, including corrections for educational status and age. The calculated risk of developing AD was 0.457, which was a reduced risk compared with the subjects who did not take ER. Interestingly, it was also determined that there was a 0.45 risk associated with taking nonsteroidal anti-inflammatory drugs. Another shorter-term study of 1124 postmenopausal women yielded similar results (Tang et al. 1996). The subjects were recruited by mail and were asked to attend an initial interview and physical examination and at least one annual follow-up examination (duration of follow-up was one to five years). At the initial interview, information on ER status was collected. At the time the data were analyzed, 12.5 percent of subjects had taken ER and 16 percent of the non-ER subjects developed AD compared with 6 percent of ER users. The calculated risk of AD for the subjects taking ER was 0.40. In addition, for the subjects taking ER who did develop AD, the onset of the disease was significantly later than in the non-ER subjects. It has also been reported from a study of 318 women with AD, that the subjects taking ER (14.5 percent) showed a significantly greater response to tacrine (an anti–Alzheimer's disease drug used at that time) therapy than the non-ER subjects (Schneider et al. 1996) (see Chapter 8).

Psychiatry

Depression and anxiety (particularly panic disorder, phobias, and obsessive-compulsive disorder) occur more frequently in females. The responses to the drugs used to treat these disorders also differ between the sexes (see Chapter 8). Neither one of these statements should be surprising to people even marginally associated with the health care system. What is surprising is that so little effort has been expended to try to understand how and why these differences occur.

Depression

Depressive disorder or "major depression" is defined in the *DSM-IV* as one or more episodes of depressed mood lasting two weeks or more involving feelings of depression and loss of interest and pleasure in all aspects of life. Such episodes are often accompanied by changes in body weight and sleep patterns, fatigue, loss of sexual drive, loss of self-esteem, and sometimes suicidal thoughts. Dysthymic disorder is defined as ongoing feelings of depression (as opposed to depressive episodes) of at least two years' duration, with feelings of depression occurring almost every day and no period greater than two months without feelings of depression. Both depressive disorder and dysthymic disorder occur approximately twice as frequently in females compared with males. There is evidence for a genetic component to depression, particularly in females.

A study of 2,662 pairs of twins (1,747 female pairs, 915 male pairs) followed the subjects over a 14-year period (Bierut et al. 1999). The twin pairs were classified as follows: female-monozygotic, n = 928 pairs; female-dizygotic, n = 527 pairs; male-monozygotic, n = 395 pairs; male-dizygotic, n = 228 pairs; female/male-dizygotic, n = 584 pairs. The pairs of twins, who volunteered for the study between 1980 and 1982, completed an initial questionnaire at the time they entered the study. Eight years later (1988–90) they completed a second questionnaire. The second questionnaire was followed by a telephone survey approximately four years later (1992–93). The results reported in the study were based on the telephone survey. The mean ages of the subjects at the time of the telephone survey were 44 years for the female subjects and 42 years for the male subjects. The survey used a modified form of the Semi-Structured Assessment for the Genetics of Alcoholism, which solicits data on psychiatric disorders as well as alcoholism. Analysis of the results showed that there was a greater concordance rate of depression in female monozygotic twins than in female dizygotic twins (Table 7.4). For male twins, the concordance rate for depression was similar for monozyotic and dizygotic twins.

A study of 1,033 pairs of female twins also examined the question of genetic and environmental influences in the development of depression

Table 7.4 Prevalence and concordance rate for *DSM-IV* Major Depressive Disorder in 2,662 pairs of twins.

Twin Pair	Prevalence	Concordance
Monozygotic-female	21.0	0.38
Dizygotic-female	24.0	0.25
Monozygotic-male	15.3	0.20
Dizygotic-male	17.3	0.23
Dizygotic-mixed:		
Female twin	24.1	0.36
Male twin	14.9	0.22

From Bierut et al. (1999).

(Kendler et al. 1992). Personal interviews with each subject were used to obtain the data (although the form of the interview was not specified). Analysis of the interview results revealed a lifetime prevalence of major depression of 31 percent and a one-month prevalence of Generalized Anxiety Disorder (GAD) of 23.5 percent. The results of this study are particularly interesting, because they suggest that genetic factors are important in the etiology of depressive disorder and that these factors also figure in the etiology of GAD. The authors suggest that the genetic factors predispose the individual to develop one of the two disorders and it is the nonfamilial environmental factors that determine which disorder will be expressed.

When the symptoms of depression are restricted to the late luteal phase of the menstrual cycle, the term "premenstrual dysphoric disorder" (PMDD) is applied. PMDD is distinguished from premenstrual syndrome (PMS) on the basis of severity. PMS symptoms are usually described in terms such as "irritability," "tension," and "depression," and do not fulfill the criteria for a major depressive episode (Yonkers 1997). Interestingly, PMDD is associated in 30 to 50 percent of sufferers with depressive disorder. There have been mixed reports on the relationship of female suicide to the menstrual cycle. Some studies have reported a higher rate in the first week; others have reported the rate to be higher in the fourth week. A study of 113 females who attempted suicide has reported that 36 percent of the attempts were made during the first week of the cycle, 19 percent in the second week, 16 percent in the third week and 29 percent in the fourth week (Baca-Garcia et al. 1998). This result suggests that there may be increased vulnerability when estrogen is low (Figure 7.8).

Bipolar disorder, characterized by episodes of depression alternating with episodes of mania (hyperactivity, elevated mood or irritability, inflated self-esteem, lasting at least a week), is represented equally in females and males. There is a subcategory of bipolar disorder, a rapid-cycling type characterized by four or more cycles per year, which is more common in females than in males. A study of 186 females and 141 males with bipolar disorder found no female-male differences in age of onset (mean age 21 years) or in living conditions (Blehar et al. 1998). The level of education attained was significantly higher for the males than the females, however. Of the female subjects, 75 percent reported having been pregnant, and of these subjects 45 percent reported experiencing severe emotional problems associated with pregnancy and birth. Thirty-one percent of the subjects reported that they had experienced menopause, 19 percent of whom reported that they had experienced severe emotional difficulties related to menopause. Finally, in relation to the menstrual cycle, 66 percent of subjects reported predictable changes in mood associated with the premenstrual or menstrual phase of the cycle (75 percent reported mood fluctuations and irritability; 25 percent reported depression) (Figure 7.8). It

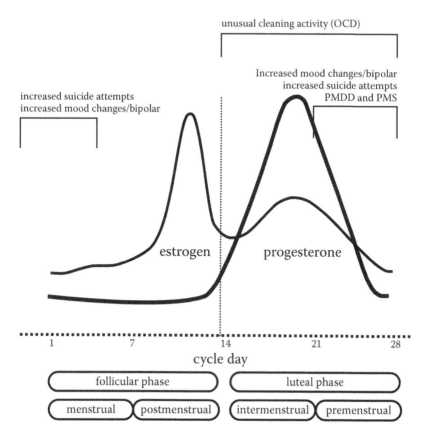

Figure 7.8 Depression, anxiety, and the menstrual cycle.

was also reported in this study that significantly more females than males in the sample had a history of hyperthyroidism and migraine.

Postpartum mood disorders usually occur in the first three months following delivery. "Postpartum blues" is an acute episode of dysphoria, tearfulness, and anxiety that occurs in the first three to seven days after delivery. It is considered by many health-care providers to be a normal part of childbirth. Postpartum depression is estimated to occur in 10 percent to 20 percent of deliveries. Predisposing factors of postpartum depression are a history of depressive disorder or a previous episode of postpartum depression itself (Pariser 1993).

Although, anecdotally, depressive disorder is associated with menopause, the relationship is not justified in the literature (Hunter 1996). In fact, for both males and females, the number of episodes of affective disorders (which includes depression) is less in the 45- to 64-year-old age group than in the 30- to 44-year-old age group (Pariser 1993). The menopause-depression association is pervasive, however, and it has been suggested that while depression is not directly associated with menopause, there may be an increased rate of recurrence for women who have already suffered from depression (Pearlstein 1995).

Anxiety

The *DSM-IV* classification for Generalized Anxiety Disorder (GAD) specifies that excessive anxiety and worry must have been experienced on an almost daily basis for a period of not less that six months. In addition to the anxiety, the person may experience restlessness, difficulty in concentrating, irritability, and sleep disturbances (and fatigue). Anxiety disorders are generally more prevalent in females than males.

GAD develops twice as frequently in females as in males. Dysthymia is more likely to develop in female GAD sufferers than in male GAD sufferers, and co-morbidity of GAD and dysthymia usually indicates a poor long-term outcome with fewer periods of remission (Pigott 1999). There is evidence to suggest that genetic factors are important in the development of GAD and that the same (or similar) genetic factors may also predispose the individual to Major Depressive Disorder (see above) (Kendler et al. 1992). GAD is also more frequent in PMS. A study of 41 GAD + PMS patients has reported that GAD symptoms were of greater severity than the symptoms of female GAD-only patients or control subjects. In the GAD + PMS patients, the GAD symptoms were also worse in the premenstrual phase (McLeod et al. 1993). In addition to GAD, there are other categories of anxiety disorders that are more prevalent in females.

Panic disorder (that is, recurrent, unexpected panic attacks) is experienced by 8 percent of females but only 3 percent of males. Panic disorder in females is also qualitatively different from the disorder in males, with females reporting more symptoms (e.g., palpitations, trembling, dizziness, etc.). There is also a greater risk in females with panic disorder for the development of agoraphobia, and for a recurrence of symptoms following remission (Yonkers et al. 1998). Anxiety may be induced in people suffering from panic disorder by CO_2 (carbon dioxide) inhalation. A study of 10 panic disorder patients and 7 healthy control subjects has reported that anxiety induced by CO_2 inhalation was significantly worse in the early-follicular phase than in the mid-luteal phase in panic disorder patients, while the control subjects experienced no anxiety reactions in either phase (Perna et al. 1995). Somatization disorder (previously referred to as hysteria), the experience of physical symptoms such as pain, gastrointestional distress, and pseudoneurological signs that cannot be fully explained by a medical condition, occurs significantly more frequently in females with panic disorder than in females without psychiatric illness (Battaglia et al. 1995).

Post-traumatic stress disorder (PTSD) is characterized by the re-experiencing of a fear-inducing, horrifying, and/or life-threatening experience. The person suffering PTSD will repeatedly relive the traumatic experience, for example, as nightmares, flashbacks, or illusions. In addition to causing distress, PTSD usually also disrupts work, family, and social life. The prevalence rate for PTSD is usually reported to be approximately the same for females and males. The cause of PTSD in females is most often assault (physical or sexual), a life-threatening experience, or being witness to a life-threatening experience (Foa 1997). PTSD in males is most often combat related. Recently, with females now being included in combat situations, there are also reports of combat-related PTSD.

A type of anxiety disorder that seems to have captured the imagination of stand-up comedians is obsessive-compulsive disorder (OCD). Stories about roommates who repeatedly check power points or stir their tea 10 times in a clockwise direction are usually good for a laugh. Lady Macbeth probably suffered from OCD. The disorder is characterized by obsessions, for example, recurrent thoughts ("Have I turned off the water in the bathtub?"), and compulsions, repeated behaviors (e.g., returning to the bathroom multiple times to check that the tap is off). Because most people indulge in checking or superstitious behaviors at some time, the behaviors must be distressing and disruptive to ongoing activities to be classified as a disorder. OCD is more frequent in females than in males, has a later onset in females, and is usually less severe (Pigott 1999). There is also evidence that in some cases OCD may be linked to the menstrual cycle. Eighteen undergraduate females (mean age 19 years) took part in a two-month study. Four of the subjects were taking an oral contraceptive. Each day, the subjects were asked to note on a questionnaire if they had completed any "usual" cleaning chores on that day. They were also asked to report if they had engaged in any "out of the ordinary cleaning activity," for example, cleaning something they do not usually clean or spending a significantly greater amount of time in cleaning activities. Fourteen incidents of "unusual cleaning behavior" were reported by 10 subjects; 11 of these 14 incidents took place during the luteal phase of the subjects' menstrual cycle (Figure 7.8). None of the subjects taking oral contraceptives reported unusual cleaning activities (Dillon and Brooks 1992).

Social phobia, also known as social anxiety disorder, is a fear of public situations where the sufferer may be subject to embarrassment (e.g., proposing a toast at a dinner party). When placed in the potentially embarrassing situation, the person experiences severe anxiety or even a panic attack (*DSM-IV*). Social phobia is more common in females than in males, and there is evidence to suggest that genetic factors are involved in its development (Stein et al. 1998).

Suicide is usually associated with depression, and given the higher incidence of depression in females than in males it is not surprising that females make more suicide attempts than males, although males have higher suicide rates (see Oquendo et al. 2007 for a review). There are numerous studies listing risk factors for suicide, and the lists tend to differ between females and males and between each other. What is consistent from the literature is that the death of a child is associated with suicide in both males and females, while marriage and children have a protective value for both sexes.

Schizophrenia

Schizophrenia is defined by the presence of psychosis (hallucinations, delusions) that has lasted for at least six months. The expression of psychosis may take the form of florid or bizarre behavior (positive signs) or withdrawal and lack of emotional expression (negative signs). Within the diagnosis of schizophrenia there are many subcategories that further define the disease (*DSM-IV*). The diagnosis of schizophrenia is not straightforward. It is heavily dependent upon cultural factors. For example, in a culture where communication with one's ancestors is an accepted part of life, hearing voices may be a cultural norm rather than an auditory hallucination. Even in similar cultural settings — for example, the United States and the United Kingdom — there are differences in the application of the diagnostic criteria. The incidence of schizophrenia is usually reported to be equal for females and males. Analysis of different populations will give somewhat different results, however. Hospital-based studies, using in-patient samples, often report that the incidence is greater in males than in females. The prevalence is estimated to be 0.5 percent to 1 percent of the general population.

Schizophrenia is usually reported to be less severe in females, with a better prognosis, although whether this difference in disease severity is consistent throughout the lifespan has been questioned (Seeman 1997). The age of onset is generally later in females (later 20s) than males (early 20s); however, in familial types of schizophrenia, the age of onset is the same for females and males (Seeman 1997). Genetic factors are clearly important in schizophrenia, the incidence of the disease in first-degree relatives of patients being 10 times the incidence in the general population. For late-onset schizophrenia, over the age of 45 years, the incidence in females is greater than in males. This female disadvantage in the late-onset group may be associated with estrogen changes (Seeman 1997). In animal models of psychosis, it has been demonstrated that estrogen, but not testosterone, reduces the behavioral abnormality of the disorder (Hafner et al. 1991).

The symptom profile in schizophrenia has often been reported to be different between females and males. Negative symptoms are generally reported to occur less frequently in females (e.g., Szymanski et al. 1995; Goldstein et al. 1990); however, some authors disagree that this is the case

(Addington et al. 1996; Seeman 1996). Paranoia has been reported to occur more frequently in females (Andia et al. 1995).

Variations from normal brain structure associated with schizophrenia are generally reported to be more pronounced in males than in females. The normal pattern of hemispheric asymmetry has often been reported (using CT and MRI) to be reversed in schizophrenia, although a number of studies have reported no differences between schizophrenia and control subjects (Bullmore et al. 1995). In many cases, differences in methodology probably account for the conflicting results. Bullmore et al. (1995) have used MRI and "radius of gyration" analysis, which they suggest overcomes many of the previous analysis problems, to examine differences in structure between the brains of patients with schizophrenia and control subjects. Radius of gyration analysis measures the way that individual points in a structure are distributed about the center of the structure. The outcome of the analysis is a value, the radius of gyration (R_g). A large R_g indicates that the points are widely dispersed about the center of the object, a small R_g indicates that the points are concentrated centrally. Thirty-seven patients with schizophrenia (8 female, 29 male) and 30 control subjects (13 female, 17 male) were examined in this study. In the right-handed male control subjects, the R_g of the right hemispheres was significantly greater than the R_g of the left hemispheres, and the pattern was reversed in the left-handed subjects. When the data were analyzed by sex, there were differences between the hemispheres for males but not for females. In right-handed males with schizophrenia (n = 26), the left-right pattern was reversed, with R_g reduced for the right hemisphere. There was no difference between the right-handed females with schizophrenia and the right-handed control females.

Bryant et al. (1999) examined 59 patients (23 female, 36 male) with schizophrenia and 37 control subjects (18 female, 19 male) using MRI. The mean duration of illness for the male and female patients was 12 years and 16 years, respectively. The areas measured in the study were the superior temporal gyrus, the amygdala/hippocampal complex, the prefrontal cortex, and the caudate nucleus. The volumes of the superior temporal gyrus and the amygdala/hippocampal complex were significantly smaller in patients compared with controls. In male patients, the left temporal lobe volume was significantly smaller than in male controls. There was no difference in total temporal lobe volume between female patients and controls. Cowell et al. (1996) used MRI to correlate changes in frontal lobe volume with the clinical signs of schizophrenia. The study included 91 people with schizophrenia (37 female, 54 male) and 114 control subjects (52 female, 62 male). The average disease duration was seven years for the females and nine years for the males. The authors found a correlation between frontal lobe changes and symptoms and that there were sex differences in these correlations (Table 7.5).

Table 7.5 Correlation between frontal lobe changes and clinical signs.

Symptom Classification	Frontal Volume: Females	Frontal Volume: Males
Negative signs	No correlation	No correlation
Disorganization	Increased volume	Decreased volume
	Increased symptoms	Increased symptoms
Hallucinations/delusions	No correlation	No correlation
Suspicion/hostility	Increased volume	No correlation
	Increased symptoms	

Source: From Cowell et al. (1996).

Sex differences have also been reported in the inferior parietal lobe volume of people with schizophrenia. Frederikse et al. (2000) conducted an MRI study of 30 people (15 female, 15 male) with schizophrenia and 30 control subjects (15 female, 15 male). The disease duration for the subjects in this study was not given. For the female patients, total brain volume was significantly less than for control females. For the male patients, there was no difference in overall brain volume between patients and controls. However, there was a significant decrease in the volume of the left inferior parietal lobe compared with controls, and this left inferior parietal lobe decrease resulted in a reversal of the usual left-right asymmetry of this region. Two postmortem anatomical studies have also reported differences between females and males. One study examined the anterior commissure of 17 brains of patients with schizophrenia (8 female, 9 male) and 20 control brains from people without a neuropsychiatric disorder (10 female, 10 male) (Highley et al. 1999). The mean disease duration was approximately 40 years for the females and 34 years for the males. The anterior commissure was measured for cross-sectional area and fiber count. The cross-section area was not different between patient and control brains; however, in the patient brains there was a significant decrease in fiber count in the female brains only. The extent of gyrification (folding) of the cortices has been suggested to be an anatomical index of schizophrenia. A second postmortem study measured the extent of gyrification of the prefrontal cortices from 24 patients with schizophrenia (13 female, 11 male) and 24 control patients (13 female, 11 male) (Vogeley et al. 2000). The average disease duration for females and males was 19 years and 25 years, respectively. There was significantly greater gyrification on the right side of the brains of the male patients compared with the male controls. There was no difference between the female patients and the female control subjects. Because the formation of the gyri is stable soon after birth, the authors suggested that this result indicated a developmental component in schizophrenia (Table 7.6).

One confounding factor in many of the anatomical studies has been that the subjects have been long-term sufferers of schizophrenia, often

Table 7.6 Anatomical changes associated with schizophrenia.

Anatomical measure	Results
Hemispheres, volume (MRI)	RH m controls, right > left
	LH m controls, left > right
	RH m patients, left > right
	RH f controls, right = left
	RH f patients, right = left
Superior temporal gyrus, amygdala/ hippocampal complex, volume (MRI)	patients < controls
Left temporal lobe, volume (MRI)	m patients < m controls
	f patients = f controls
Lateral ventricles, volume (CT)	patients > controls
Lateral ventricle to brain ratio	f patients > m patients
Total brain volume (MRI)	f patients < f controls
	m patients = m controls
Inferior parietal lobe, volume to brain ratio (MRI)	L, m patients < m controls
	f patients = f controls
Anterior commissure, area and fiber count (post-mortem)	Area: patients = controls
	Count: f patients < f controls
Gyrification of prefrontal cortices (post-mortem)	R, m patients > m controls
	f patients = f controls

RH, right-handed; LH, left-handed; m, male; f, female; R, right side of the brain; L, left side of the brain.

hospitalized, and with a long history of treatment with antipsychotic medications. In many cases, the drug history (and sometimes the disease duration) is either unknown or is not included in the study, making interpretation of the results difficult. It is impossible to determine if the observed changes are due to the disease or some other factor — for example, drug treatments. Ideally, data should be collected before drug treatment has commenced. A CT study of people experiencing their first episode of schizophrenia has reported that the lateral ventricles of the patients (n = 63) was larger than in controls (n = 21), resulting in a larger ventricle to brain ratio in the patients. This ratio was larger in the female patients (n = 30) than in the male patients (n = 33) (Vazquez-Barquero et al. 1995). An MRI study of the hippocampus-amygdala and temporal horns in first episode schizophrenia (n = 34) compared with healthy control subjects (n = 25) has also yielded significant results (Bogerts et al. 1990). This study reported that, in male (n = 22) and female patients (n = 12), there was an enlargement of the temporal horn on the left side compared with controls. There was also a significant decrease in the volume of the left hemisphere of male patients compared with control subjects.

A number of cognitive changes have been reported to occur in schizophrenia. The greatest cognitive impairments have generally been reported in males. It has also been suggested that impairment is greatest for left-hemisphere tasks and that disruptions in the patterns of laterality of brain function occur in schizophrenia. A 1997 study by Reite et al. demonstrated that the auditory-evoked field component of magneto-encephalography (MEG) is disrupted in schizophrenia. The responses of a group of 20 patients with paranoid schizophrenia (9 female, 11 male) and 20 control subjects (10 female, 10 male) were compared using MRI and MEG. As discussed in Chapter 5, the M100 source in the male control subjects was more asymmetrical than in the female control subjects. In the male patients, the M100 source was significantly less asymmetrical than in the control subjects, but in the female patients there was greater asymmetry than in the female control subjects. The superior temporal gyri (location of the M100 source) were also smaller in the male patients than in the male control subjects (Reite et al. 1997).

Gruzelier et al. (1999) tested the responses of people with schizophrenia on a memory task for unfamiliar faces and words. There were 104 patients in the study (38 female, 66 male) and 95 control subjects (50 female, 45 male). The patients were categorized according to symptoms (delusional, positive symptoms, negative symptoms) and their disease state (active [7 female, 17 male], withdrawn [5 female, 11 male], mixed [12 female, 20 male], and remitted [14 female, 18 male]). Medication analysis by disease state was also given. The study was conducted in two stages. In the learning stage, 50 items were presented for 3 seconds each and the subjects were required to give an oral response (to ensure attention was focused on the task) indicating whether they "liked" or "disliked" the item presented. In the recognition stage, 50 pairs of items were presented; one item of each pair had been presented in the learning condition. The subjects were required to indicate the previously presented item. Overall, the control subjects demonstrated greater accuracy than the patients, and males were more accurate than females. The female control subjects were more accurate for words than faces and the male controls were more accurate for faces. Within the patient groups, the males in the active category were more accurate for words than faces (a reversal of control performance). Both males and females in the withdrawn group were more accurate for faces than words (a reversal from controls for the females). For subjects in remission, females performed more accurately than males. These results are generally consistent with the commonly held view that left hemisphere function is more impaired than right hemisphere function in schizophrenia. These results are not consistent with the view that females fare better than males. However, the study was of hospitalized patients (except for some patients in the remission group), so it may be that the female subjects represented a subset of more seriously affected females.

A study of 75 patients with schizophrenia (30 female, 45 male) and 75 control subjects (30 female, 45 male) used a battery of 18 tests to assess cognitive skills (Ragland et al. 1999). The patients and controls were all right-handed and age-matched. The average disease duration for the patients was 10 years. The tests included measures of the following abilities: abstraction, attention, verbal memory, spatial memory, language ability, spatial (line orientation, block design), sensory (stereognosis, i.e., identification of objects by touch alone), and motor function (finger tapping). On the finger-tapping test, both patients and controls tapped significantly faster with their right hands, and in both groups males tapped faster than females. When asked to identify objects by touch alone, control subjects were faster than the patients. The control subjects were also faster with their left hands, but right- and left-hand times were equal for the patients. On language and spatial ability, the patients performed better on spatial tasks than language tasks, and males performed better on the spatial tasks. Female control subjects performed better on the verbal tasks than on the spatial tasks, but female patients performed equally well on both tasks. Comparing spatial and verbal memory, control subjects performed better than patients, but there were no sex differences in performance. The authors noted that while greater left-hemisphere impairment may hold for some tasks, at least in the present study, it was not apparent for the motor and sensory tasks.

Variations in symptom severity across the menstrual cycle have been reported in schizophrenia. A group of 32 in-patient females with a mean age of 31 years was assessed on days 2, 7, 13, 14, 21, and 28 of their menstrual cycles (Riecher-Rössler et al. 1994). The assessment on each of these days included psychiatric symptoms, self-assessed well-being, and analysis of serum levels of estrogen and progesterone. In all of the patients there was a significant overall reduction in estrogen levels as compared with population norms. This overall reduction in estrogen resulted in a "flattening" of the estrogen level changes across the cycle. However, there was still an effect of estrogen on psychiatric state, and self-assessment, with the fewest symptoms and the greatest well-being reported when estrogen levels were highest. Similar results were reported for a study of 39 female in-patients, mean age 35 years (Harris 1997). The patients were assessed in the premenstrual phase (days 23–28) and the postmenstrual phase (days 5–10), using similar measures to the previous study. Analysis of the results revealed that fewer symptoms were reported when estrogen was elevated. Interestingly, the increased symptoms reported when estrogen was low were related to affective disorder (e.g., depression) rather than psychosis. A study of five outpatients with schizophrenia also reported a significant increase in symptoms associated with low estrogen levels (Hallonquist et

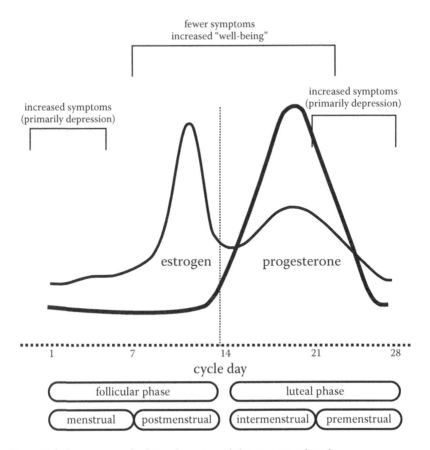

fewer symptoms
increased "well-being"

increased symptoms
(primarily depression)

increased symptoms
(primarily depression)

estrogen   progesterone

1   7   14   21   28

cycle day

follicular phase   luteal phase

menstrual   postmenstrual   intermenstrual   premenstrual

Figure 7.9 Symptoms of schizophrenia and the menstrual cycle.

al. 1993). Interestingly, the symptom that figured most prominently (and changed the most) was depression (Figure 7.9).

Finally, there is also evidence for changes in the symptoms associated with childbirth and the postpartum period, but the results are inconsistent.

Conclusions

The disorders discussed in this chapter represent the most prominent neuropathologies facing the twenty-first-century population. In some cases, for example, Alzheimer's disease and Parkinson's disease, the increasing incidence is a by-product of our increasing life expectancy. In the case of disorders such as depression, anxiety, and schizophrenia, it is our awareness and understanding that has increased. It is interesting that the majority of these disorders are more prevalent in females than in males (Table 7.7). The recognition of the vulnerability of females to some

Table 7.7 Prevalence of Neuropathologies in Females and Males

Disorder	Prevalence
Alzheimer's disease	F > M
Amyotrophic lateral sclerosis	F < M
	F = M*
Anxiety	F > M
Bipolar disorder	F = M
Depression	F > M
Epilepsy	F = M
Migraine	F > M
Multiple sclerosis	F > M
Obsessive compulsive disorder	F > M
Panic disorder	F > M
Parkinson's disease	F < M
Post-traumatic stress disorder	F = M
Schizophrenia	F = M
Social phobia	F > M
Substance abuse	F < M

*Data for postmenopausal females with age of symptom onset over 60 yrs.

neurological and psychiatric disorders is slowly changing the perspective of health-care providers and others concerned with female health-care issues. For the interested reader, *Folding Back the Shadows* (Romans 1998) provides an interesting and challenging perspective on these issues.

The following summarizes the main points to be drawn from Chapter 7:

1. Overall, females appear to be more vulnerable to the development of the most common neurological and psychological disorders.
2. In the majority of cases, an important hormone seems to be estrogen. We will return to this subject in Chapter 8 in the discussion of drugs and their actions in females. The exception may be MS, for which there is some evidence to support a beneficial role for progesterone.
3. For a number of the disorders, menstrual-cycle changes have also been demonstrated. Again, the majority of symptoms seem to worsen during the low estrogen phases of the cycle. The exception is MS, unless one looks at changes in estrogen as opposed to absolute values.

Bibliography and recommended readings

Abplanalp, J. M., L. Livingston, R. M. Rose, and D. Sandwisch. 1977. Cortisol and growth hormone responses to psychological stress during the menstrual cycle. *Psychosomatic Medicine* 39: 158–77.

Addington, D., J. Addington, and S. Patten. 1996. Gender and affect in schizophrenia. *Canadian Journal of Psychiatry* 41: 265–68.

Andia, A. M., S. Zisook, R. K. Heaton, J. Hesselink, T. Jernigan, J. Kuck, J. Morganville, and D. L. Braff. 1995. Gender differences in schizophrenia. *Journal of Nervous and Mental Disorders* 183: 522–28.

Arnason, B. G., and D. P. Richman. 1969. Effect of oral contraceptives on experimental demyelinating disease. *Archives of Neurology* 21: 103–08.

Baca-Garcia, E., A. Sanchez Gonzalez, P. Diaz-Corralero, I. Bonzalez Garcia, and J. de Leon. 1998. Menstrual cycle and profiles of suicidal behaviour. *Acta Psychiatrica Scandinavica* 97: 32–35.

Bansil, S., H. J. Lee, S. Jindal, C. R. Holtz, and S. D. Cook. 1999. Correlation between sex hormones and magnetic resonance imaging lesions in multiple sclerosis. *Acta Neurolgica Scandanavica* 99: 91–94.

Barrett, K. M., T. G. Brott, R. D. Brown Jr., M. R. Frankel, B. Bradford, B. B. Worrall, S. L. Silliman, L. D. Case, S. S. Rich, and J. F. Meschia. For the Ischemic Stroke Genetics Study Group. 2007. *Journal of Stroke and Cerebrovascular Diseases* 16: 34–39.

Battaglia, M., L. Bernardeschi, E. Politi, S. Bertella, and L. Bellodi. 1995. Cormorbidity of panic and somatization disorder; a genetic-epidemiological approach. *Comprehensive Psychiatry* 36: 411–20.

Behl, C., M. Widmann, T. Trapp, and F. Holsboer. 1995. 17-β estradiol protects neurons from oxidative stress-induced cell death *in vitro*. *Biochemical and Biophysical Research Communications* 216: 473–82.

Bierut, L. J., A. C. Heath, K. K. Bucholz, S. H. Dinwiddie, P.A.F. Madden, D. J. Statham, M. P. Dunne, and N. G. Martin. 1999. Major depressive disorder in a community-based twin sample. *Archives of General Psychiatry* 56: 557–63.

Birk, K., C. Ford, S. Smeltzer, D. Ryan, R. Miller, and R. A. Rudick. 1990. The clinical course of multiple sclerosis during pregnancy and the puerperium. *Archives of Neurology* 47: 738–42.

Blehar, M. C., J. R. DePaulo Jr., E. S. Gershon, T. Reich, S. G. Simson, and J. I. Nurnberger Jr. 1998. Women with bipolar disorder: Findings from the NIMH genetics initiative sample. *Psychopharmacology Bulletin* 34: 239–43.

Bogerts, B., M. Ashtari, G. Degreef, J. M. Alvir, R. M. Bilder, and J. A. Lieberman. 1990. Reduced temporal limbic structure volume on magnetic resonance images in first episode schizophrenia. *Psychiatry Research: Neuroimaging* 35: 1–13.

Bousser, M-G. 2004. Estrogens, migraine and stroke. *Stroke* 35 (Suppl. 1): 2652–56.

Bowman, R. E. 2005. Stress-induced changes in spatial memory are sexually differentiated and vary across the lifespan. *Journal of Neuroendocrinology* 17: 526–35.

Bromberg, M. B. 1999. Pathogenesis of amyotrophic lateral sclerosis: A critical review. *Current Opinion in Neurology* 12: 581–88.

Bryant, N. L., R. W. Buchanan, K. Vladar, A. Breier, and M. Rothman. 1999. Gender differences in temporal lobe structures of patients with schizophrenia: A volumetric MRI study. *American Journal of Psychiatry* 156: 603–09.

Bullmore, E., M. Brammer, I. Harvey, R. Murray, and M. Ron. 1995. Cerebral hemispheric asymmetry revisited; effects of handedness, gender and schizophrenia measured by radius of gyration in magnetic resonance images. *Psychological Medicine* 25: 349–63.

Confavreux, C., M. Hutchinson, M. M. Hours, P. Cortinovis-Tourniaire, T. Moreau, and the Pregnancy in Multiple Sclerosis Group. 1998. Rate of pregnancy-related relapse in multiple sclerosis. *New England Journal of Medicine* 339: 285–91.

Conway, C. A., B. C. Jones, L. M. DeBruine, L. L. Welling, M. J. Wal Smith, D. I. Perrett, M. A. Sharp, and E. A. Al-Dujaili. 2007. Salience of emotional displays of danger and contagion in faces is enhanced when progesterone levels are raised. *Hormones and Behaviour* 51: 202–06.

Cowell, P. E., D. J. Kostianovsky, R. C. Gur, B. I. Turetsky, and R. E. Gur. 1996. Sex differences in neuroanatomical and clinical correlations in schizophrenia. *American Journal of Psychiatry* 153: 799–805.

Coyle, D. E., C. S. Sehlhorst, and M. M. Behbehani. 1996. Intact female rats are more susceptible to the development of tactile allodynia than ovariectomized female rats following partial sciatic nerve ligation (PSNL). *Neuroscience Letters* 203: 37–40.

Coyle, D. E., C. S. Sehlhorst, and C. Mascari. 1995. Female rats are more susceptible to the development of neuropathic pain using the partial sciatic nerve ligation (PSNL) model. *Neuroscience Letters* 186: 135–38.

Dao, T. T. T., K. Knight, and V. Ton-That. 1998. Modulation of myofascial pain by the reproductive hormones: A preliminary report. *Journal of Prosthetic Dentistry* 79: 663–70.

Diamond, S. G., C. H. Markham, M. M. Hoehn, F. H. McDowell, and M. D. Muenter. 1990. An examination of male-female differences in progression and mortality of Parkinson's disease. *Neurology* 40: 763–66.

Dillon, K. M., and D. Brooks. 1992. Unusual cleaning behavior in the luteal phase. *Psychological Reports* 70: 35–39.

DSM-IV. 1994. *Diagnostic and Statistical Manual of the American Psychiatric Association*. 4th ed. Washington, D.C : American Psychiatric Association.

Dzaja, A., S. Arber, J. Hislop, M. Kerkhofs, C. Kopp, T. Pollmacher, P. Polo-Kantola, D. J. Skene, P. Stenuit, I. Tobler, and T. Porkka-Heiskanen. 2005. Women's sleep in health and disease. *Journal of Psychiatric Research* 39: 55–76.

Erstvag, J. M., J-A. Zwart, G. Helde, H-J. Johnsen, and G. Bovim. 2005. Headache and transient focal neurological symptoms during pregnancy, a prospective cohort. *Acta Neurologica Scandanavica* 111: 233–37.

Foa, E. B. (1997) Trauma and women: Course, predictors, and treatment. *Journal of Clinical Psychiatry* 58 (Suppl 9), 25–28.

Fox, H. C., K. A. Hong, P. Paliwal, P. T. Morgan, and R. Sinha. 2008. Altered levels of sex and stress steroid hormones assessed daily over a 28-day cycle in early abstinent cocaine-dependent females. *Psychopharmacology* 195: 527–36.

Frederikse, M., A. Lu, E. Aylward, P. Barta, T. Sharma, and G. Pearlson. 2000. Sex differences in inferior parietal lobule volume in schizophrenia. *American Journal of Psychiatry* 157: 422–27.

Gilmore, W., L. P. Weiner, and J. Correale. 1997. Effect of estradiol on cytokine secretion by proteolipid protein-specific T cell clones isolated from multiple sclerosis patients and normal control subjects. *Journal of Immunology* 158: 446–51.

Glover, V., T. G. O'Connor, J. Heron, and J. Golding. ALSPAC Study Team. 2004. Antenatal maternal anxiety is linked with atypical handedness in the child. *Early Human Development* 79: 107–18.

Golbe, L. 1987. Parkinson's disease and pregnancy. *Neurology* 37: 1245–49.

Goldstein, J., S. L. Santangelo, J. C. Simpson, and M. T. Tsuang. 1990. The role of gender in identifying subtypes of schizophrenia; a latent class of analytic approach. *Schizophrenia Bulletin* 16: 263–75.

Gruzelier, J. H., L. Wilson, D. Liddiard, E. Peters, and L. Pusavat. 1999. Cognitive asymmetry patterns in schizophrenia: Active and withdrawn syndromes and sex differences as moderators. *Schizophrenia Bulletin* 25: 349–62.

Gutteling, B. M., C. De Weerth, and J. K. Buitelaar. 2007. Prenatal stress and mixed-handedness. *Pediatric Research* 62: 586–590.

Hafner, H., S. Behrens, J. De Vry, and W. F. Gattaz. 1991. An animal model for the effects of estradiol on dopamine-mediated behavior: Implications for sex differences in schizophrenia. *Psychiatry Research* 38: 125–34.

Hallonquist, J. D., M. V. Seeman, M. Lang, and N. A. Rector. 1993. Variation in symptom severity over the menstrual cycle of schizophrenics. *Biological Psychiatry* 33: 207–09.

Harris, A. H. 1997. Menstrually related symptom changes in women with schizophrenia. *Schizophrenia Research* 27: 93–99.

Headache Classification Committee of the International Headache Society. 1988. Classification and diagnostic criteria for headache disorders, cranial neuralgias and facial pain. *Cephalagia* 8: 1–96.

Hedmann, C., T. Pohjasvaara, U. Tolonen, J. E. Suhonen-Malm, and V. V. Myllyla. 2002. Effects of pregnancy on mothers' sleep. *Sleep Medicine* 3: 37–42.

Highley, J. R., M. M. Esiri, B. McDonald, H. C. Roberts, M. A. Walker, and T. J. Crow. 1999. The size and fiber composition of the anterior commissure with respect to gender and schizophrenia. *Biological Psychiatry* 45: 1120–27.

Hunter, M. S. 1996. Depression and the menopause. *British Medical Journal* 313: 1217–17.

Jaffe, A. B., C. D. Toran-Allerand, P. Greengard, and S. E. Gandy. 1994. Estrogen regulates metabolism of Alzheimer amyloid β-precursor protein. *Journal of Biological Chemistry* 269: 13065–68.

Kawas, C., S. Resnick, A. Morrison, R. Brookmeyer, M. Corrada, A. Zonderman, C. Bacal, D. D. Lingle, and E. Metter. 1997. A prospective study of estrogen replacement therapy and the risk of developing Alzheimer's disease: The Baltimore Longitudinal Study of Aging. *Neurology* 48: 1517–21.

Kendler, K. S., M. C. Neale, R. C. Kessler, A. C. Heath, and L. J. Eaves. 1992. Major depression and generalized anxiety disorder: Same genes, (partly) different environments? *Archives of General Psychiatry* 49: 716–22.

Marder, K., M.-X., Tang, B. Alfaro, H. Mejia, L. Cote, D. Jacobs, Y. Stern, M. Sano, and R. Mayeux. 1998. Postmenopausal estrogen use and Parkinson's disease with and without dementia. *Neurology* 50: 1141–43.

McLeod, D. R., R. Hoehn-Saric, G. V. Foster, and P. A. Hipsley. 1993. The influence of premenstrual syndrome on ratings of anxiety in women with generalized anxiety disorder. *Acta Psychiatrica Scandinavica* 88: 248–51.

Mennes, M., P. Stiers, L. Lagae, B. Van den Bergh. 2006. Long-term cognitive sequelae of antenatal maternal anxiety: Involvement of the orbitofrontal cortex. *Neuroscience and Behavioral Reviews* 30: 1078–86.

Morrell, M. J. 1999. Epilepsy in women: The science of why it is special. *Neurology* 53 (Suppl. 1): S42–S48.

Oquendo, M. A., M. E. Bongiovi-Garcia, H. Galfalvy, P. H. Goldberg, M. F. Grunebaum, A. Burke, A., and J. J. Mann. 2007. Sex differences in clinical predictors of suicidal acts after major depression: A prospective study. *American Journal of Psychiatry* 164: 134–41.

Parent, A., F. Sato, Y. Wu, J. Gauthier, M. Lévesque, and M. Parent. 2000. Organization of the basal ganglia: The importance of axonal collateralization. *Trends in Neuroscience* 23 (Suppl.): S20–S27.

Pariser, S. F. 1993. Women and mood disorders: Menarche to menopause. *Annals of Clinical Psychiatry* 5: 249–54.

Pearlstein, T. B. 1995. Hormones and depression: What are the facts about premenstrual syndrome, menopause, and hormone replacement therapy? *American Journal of Obstetrics and Gynecology* 173: 646–53.

Perna, G., F. Brambilla, C. Arancio, and L. Bellodi. 1995. Menstrual cycle-related sensitivity to 35% CO_2 in panic patients. *Biological Psychiatry* 37: 528–32.

Piccinni, M-P., M-G. Giudizi, R. Biagiotti, L. Beloni, L. Giannarini, S. Sampognaro, P. Parronchi, R. Manetti, F. Annunziato, C. Livi, S. Romagnani, and E. Maggi. 1995. Progesterone favors the development of human T helper cells producing Th2-Type cytokines and promotes both IL-4 production and membrane CD30 expression in established Th1 cell clones. *Journal of Immunology* 155: 128–33.

Piggot, T. A. 1999. Gender differences in the epidemiology and treatment of anxiety disorders. *Journal of Clinical Psychiatry* 60 (Suppl. 18): 4–15.

Pozzilli, C., P. Falaschi, C. Mainero, A. Martocchia, R. D'Urso, A. Proietti, M. Frontoni, S. Bastianello, and M. Filippi. 1999. MRI in multiple sclerosis during the menstrual cycle: Relationship with sex hormone patterns. *Neurology* 53: 622–24.

Ragland, J. D., R. E. Gur, B. C. Klimas, N. McGrady, and R. C. Gur. 1999. Neuropsychological laterality indices of schizophrenia: Interactions with gender. *Schizophrenia Bulletin* 25: 79–89.

Reite, M., J. Sheeder, P. Teale, M. Adams, D. Richardson, J. Simon, R. H. Jones, and D. C. Rojas. 1997. Magnetic source imaging evidence of sex differences in cerebral lateralization in schizophrenia. *Archives of General Psychiatry* 54: 433–40.

Riecher-Rössler, A., H. Häfner, M. Stumbaum, K. Maurer, and R. Schmidt. 1994. Can estradiol modulate schizophrenic symptomatology? *Schizophrenia Bulletin* 20: 203–14.

Riley, J. L. III, M. E. Robinson, E. A. Wise, and D. D. Price. 1999. A meta-analytic review of pain perception across the menstrual cycle. *Pain* 81: 225–35.

Rollman, G. B. 1993. Sex differences and biological rhythms affecting pain responsiveness. *Pain* 55: 277.

Romans, S. E., ed. 1998. *Folding Back the Shadows*. Dunedin: University of Otago Press.

Rubin, S. M. 2007. Parkinson's disease in women. *Disability Monthly* 53: 206–13.

Sadovnick, A. D., D. Bulman, and G. C. Ebers. 1991. Parent-child concordance in multiple sclerosis. *Annals of Neurology* 29: 252–55.

Saunders-Pullman, R., J. Gordon-Elliott, M. Parides, S. Fahn, H. R. Saunders, and S. Bressman. 1999. The effect of estrogen replacement on early Parkinson's disease. *Neurology* 52: 1417–21.

Savic, I., and J. Engel Jr. 1998. Sex differences in patients with mesial temporal lobe epilepsy. *Journal of Neurology, Neurosurgery and Psychiatry* 65: 910–12.

Schneider, L. S., M. R. Farlow, V. W. Henderson, and J. M. Pogoda. 1996. Effects of estrogen replacement therapy on response to tacrine in patients with Alzheimer's disease. *Neurology* 46: 1580–84.

Seeman, M. V. 1996. Schizophrenia, gender, and affect. *Canadian Journal of Psychiatry* 41: 263–64.

———. 1997. Psychopathology in women and men: Focus on female hormones. *American Journal of Psychiatry* 154: 1641–47.

Smith, R., and J. W. W. Studd. 1992. A pilot study of the effect upon multiple sclerosis of the menopause, hormone replacement therapy and the menstrual cycle. *Journal of the Royal Society of Medicine* 85: 612–13.

Spector, S., C. Cull, and L. H. Goldstein. 2000. Seizure precipitants and perceived self-control of seizures in adults with poorly-controlled epilepsy. *Epilepsy Research* 38: 207–16.

Stein, M. B., M. J. Chartier, A. L. Hazen, M. V. Kozak, M. E. Tancer, S. Lander, P. Furer, K. Chubaty, and J. R. Walker. 1998. A direct-interview family study of generalized social phobia. *American Journal of Psychiatry* 155: 90–97.

Szymanski, S., J. A. Lieberman, J. M. Alvir, D. Mayerhoff, A. Loebel, S. Geisler, M. Chakos, A. Koreen, D. Jody, J. Kane, et al. 1995. Gender differences in onset of illness, treatment response, course and biologic indexes in first-episode schizophrenic patients. *American journal of Psychiatry* 152: 698–703.

Tang, M-X., D. Jacobs, Y. Stern, K. Marder, P. Schofield, B. Gurland, H. Andrews, and R. Mayeux. 1996. Effect of oestrogen during menopause on risk and age at onset of Alzheimer's disease. *Lancet* 348: 429–32.

Thulin, P. C., J. H. Carter, M. D. Nichols, M. Kurth, and J. G. Nult. 1996. Menstrual-cycle related changes in Parkinson's disease. *Neurology* 46: A376.

Van Walderveen, M. A. A., M. W. Tas, F. Barkhof, C. H. Polman, S. T. F. M. Frequin, O. R. Hommes, and J. Valk. 1994. Magnetic resonance evaluation of disease activity during pregnancy in multiple sclerosis. *Neurology* 44: 327–29.

Vazquez-Barquero, J. L., M. J. Cuesta Nuñez, F. Quintana Pando, M. de la Varga, S. Herrera Castanedo, and G. Dunn. 1995. Structural abnormalities of the brain in schizophrenia: Sex differences in the Cantabria First Episode of Schizophrenia Study. *Psychological Medicine* 25: 1247–57.

Vogeley, K., T. Schneider-Axmann, U. Pfeiffer, R. Tepest, T. A. Bayer, B. Bogerts, W. G. Honer, and P. Falkai. 2000. Disturbed gyrification of the prefrontal region in male schizophrenic patients: A morphometric postmortem study. *American Journal of Psychiatry* 157: 34–39.

Voskuhl, R. R., H. Pitchekian-Halabi, A. MacKenzie-Graham, H. F. McFarland, and C. S. Raine. 1996. Gender differences in autoimmune demyelination in the mouse: Implications for multiple sclerosis. *Annals of Neurology* 39: 724–33.

Woolley, C. S., and P. A. Schwartzkroin. 1998. Hormonal effects on the brain. *Epilepsia* 39 (Suppl. 8): S2–S8.

Yaffe, K., K. Lindquist, S. Sen, J. Cauley, R. Ferrell, B. Penninx, T. Harris, R. L. K. Li, and S. R. Cummings, for the Health ABC Study (in press). Estrogen receptor genotype and risk of cognitive impairment in elders: Finding from the Health ABC Study. *Neurobiology of Aging*.

Yonkers, K. A. 1997. The association between premenstrual dysphoric disorder and other mood disorders. *Journal of Clinical Psychiatry* 58 (Suppl. 15): 19–25.

Yonkers, K. A., C. Zlotnick, J. Allsworth, M. Warshaw, T. Shea, and M. B. Keller. 1998. Is the course of panic disorder the same in women and men? *American Journal of Psychiatry* 155: 596–602.

Zahn, C. 1999. Catamenial epilepsy: Clinical aspects. *Neurology* 53 (Suppl. 1): S34–S37.

Zorgdrager, A., and J. De Keyser. 1997. Menstrually related worsening of symptoms in multiple sclerosis. *Journal of Neurological Sciences* 149: 95–97.

chapter 8

Drugs and drug effects

In many ways this chapter, which is about pharmacology, is a continuation of the last chapter. In Chapter 7 we discussed neurological and psychiatric disorders. In this chapter we will look at drug treatments for those disorders and how the drugs' actions may differ in females and males. We will also look at drugs used for non-therapeutic purposes.

Although this chapter is primarily concerned with drug actions in the brain (neuropharmacology and psychopharmacology), only in very rare instances are drugs delivered directly into the brain. Instead, drugs are usually administered orally or by injection and are carried to the brain by the blood. Along the way a number of processes take place that may alter not only the amount of drug reaching the brain, but also the form of the drug itself. These processes are part of the subject of *pharmacokinetics*, "what the body does to the drug."

Pharmacokinetic processes operate from the moment the drug enters the body to the time that it (or its by-products) leaves (Figure 8.1). The drug must be absorbed into the bloodstream from the site of administration, for example, the gastrointestinal tract after oral administration, and distributed throughout the body. Drugs absorbed from the gastrointestinal tract travel via the portal vein directly to the liver where they may undergo preliminary processing, known as *first pass metabolism*. Some preliminary processing may also occur in the intestinal epithelial cells. Some drugs undergo such extensive first pass metabolism that only a small proportion of the initial dose remains. An example of this occurs with the anxiolytic drug diazepam. Approximately 80 percent of orally administered diazepam is lost due to first pass metabolism. From the liver, the remaining drug is distributed throughout the body by the blood. The phrase "throughout the body" is particularly important. Drugs destined for a particular target are not distributed to that area alone. Once a drug enters the bloodstream it is distributed to every place that the blood goes, and if receptors for the drug are available in a particular place, then the drug will have an effect, whether such action is desirable or not. This is an important issue for drugs targeted at the CNS; many of them have unwanted side effects due to actions on receptors in other parts of the body. L-dopa, for example, the mainstay of treatment for Parkinson's disease, is metabolized to dopamine in the brain and binds to dopamine receptors. There are also dopamine receptors in the *area postrema*, the body's "emetic center," which is outside the blood brain barrier. If L-dopa is metabolized outside the

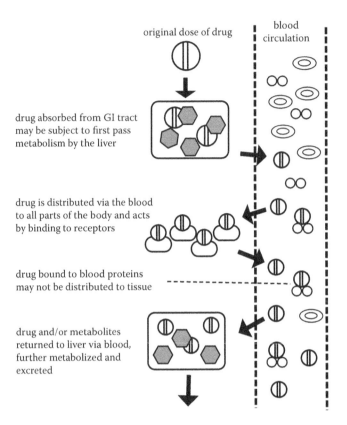

Figure 8.1 The steps in drug absorption, distribution, and elimination (pharmacokinetics) following oral administration.

blood-brain barrier, it binds to the dopamine receptors in the area postrema, producing severe nausea and vomiting. To overcome this problem, L-dopa is administered with an agent (a decarboxylase inhibitor) that prevents its metabolism outside the blood-brain barrier.

Following their distribution and (hopefully) therapeutic action, drugs or their metabolites are transported to the liver and/or the kidneys for elimination. The metabolites of some drugs also have drug actions (active metabolites) that may prolong the action of the parent drug. Diazepam, for example, is metabolized to nordiazepam (active metabolite), which is in turn metabolized to oxazepam (active metabolite) before finally being inactivated. Various factors including age, state of health, genetics, and the sex of the person taking the drug can all affect pharmacokinetics.

Pharmacodynamics is the area of pharmacology devoted to the action of drugs on receptors ("what the drug does to the body"), and it can also be altered by a number of factors. The effect that the drug has on a particular person, at a particular time, will depend upon the age and health of the person, genetic factors, previous drug exposure, environmental factors, and the sex of the person taking the drug.

There is a limited amount of data to suggest that there are differences in the pharmacokinetics and pharmacodynamics of some drugs administered to human females and males. The reason that there is a limited amount of data is that females have been excluded from participation in drug trials. The data have simply not been collected. There is, however, a great deal of evidence from the literature using experimental animals, around 60 years' worth. By necessity much of the discussion in this chapter, on CNS drug effects, as well as pharmacokinetics and pharmacodynamics, will center on experimental data from animal studies.

Pharmacokinetics and pharmacodynamics

Pharmacokinetics can be divided into its component processes: absorption, distribution, metabolism, and excretion. Each component process may be modified by a number of factors. There are a number of physiological variables that differ between females and males. Some of these variables are likely to have a direct influence on the different pharmacokinetic processes. One very important factor to be considered in the case of distribution is the difference in the proportion of body fat between females and males. In young adult females, the proportion of body fat is approximately 33 percent. This percentage rises to around 48 percent in elderly females. The proportion in males is approximately 18 percent, rising to 36 percent (Pollock 1997). The volume of distribution is a measure of the amount of drug distributed into body tissue, particularly fat. Taken on its own, how the drug is distributed into body tissue may not seem particularly interesting. However, a particularly important variable, plasma half-life (a measure of how long a drug remains in the body), is directly proportional to the volume of distribution. So, in many cases, bodies with a higher proportion of fat will retain certain drugs for a longer period of time than bodies with a lower proportion of fat. Enter sex differences in pharmacokinetics. It has been reported that the volume of distribution of alcohol is smaller in females than in males, while the volume of distribution of diazepam is larger in females than in males (Harris et al. 1995). It has also been reported that both the volume of distribution and the plasma half-life of the antidepressant drug buproprion are significantly increased in elderly females when compared with young males (Sweet et al. 1995). In the gastrointestinal tract, the rate of gastric emptying is decreased in females compared to males. The decreased rate may be observed during menopause in females not taking hormone supplements, suggesting that the slowing is not due to direct effects of estrogen or progesterone (Pollock 1997). Differences in drug metabolism may be due in part to different amounts of available liver enzymes. It has been reported, for example, that alcohol dehydrogenase activity is less in females, while CYP3A4 activity (see below) is increased in young adult females compared with males and postmenopausal females (Pollock 1997).

The next question, of course, is do these variables change across the menstrual cycle? Kashuba and Nafziger (1998) have reviewed the data on changes in a number of physiological variables associated with the menstrual cycle (that may have the potential to influence pharmacokinetics and pharmacodynamics) (Table 8.1) and the data on pharmacokinetics across the menstrual cycle. The opening paragraph of the review gives an accurate summary of the current state of this area of pharmacological research, "There is an increasing awareness that the exclusion of women from clinical trials may lead to inaccurate application of drug therapy in women. Gender and estrous cycle differences in pharmacokinetics and pharmacodynamics of drugs in animals have been appreciated for over 60 years, but investigation into these differences in humans has only recently occurred" (1998, 203).

Table 8.1 Physiological variables and the menstrual cycle.

	Cycle/hormone-related changes
Kidney function	
Creatinine clearance	No clear evidence
Glomerular filtration	No effect of estrogen
Vasopressin	Higher in luteal phase
Sodium excretion	Decreases with increasing estrogen
GI System	
Colonic transit time	f > m, no cycle differences
Cardiovascular System	
Heart rate	Higher in luteal phase
Diastolic BP	Lower in luteal phase
Systolic BP	Higher in luteal phase
Stroke volume	Increased in luteal phase
Cardiac output	Increased in luteal phase
Lipids/lipoproteins	No change across cycle
Metabolism	
Temperature*	Increase early luteal phase
Energy expenditure	Increased in luteal phase
Blood	
Platelet aggregation	Increased luteal phase
Immune responses	Decreased with high estrogen/progesterone

Source: From Kashuba and Nafziger (1998). BP, blood pressure; GI, gastrointestinal.
*Both core and skin temperature

Table 8.2 Data on pharmacokinetics and the menstrual cycle.

	Available data
Absorption	No clear evidence for cyclic changes
Distribution	Few studies, no clear evidence for changes
Metabolism	Evidence for differences in metabolism of some drugs
Excretion	Conflicting data, no clear evidence for changes

Source: From Kashuba and Nafziger (1998).

From the data on physiological variables summarized in Table 8.1, it is clear that there are functional differences between males and females that have the potential to change pharmacokinetic and pharmacodynamic parameters. One factor not included in this table is the menstrual cycle–related fluid retention that occurs in a subset of females and has the potential to lower plasma concentrations of drugs (Yonkers et al. 1992).

The only area of pharmacokinetics where there is evidence for differences associated with the menstrual cycle is in drug metabolism by the liver, sometimes referred to as biotransformation. Interestingly, the metabolism of a number of drugs is influenced by administration of oral contraceptives. The liver enzyme system involved in the metabolism of many drugs is the cytochrome P450 (CYP) system and within this system there are different isoforms of CYP with specificity for different drugs. One of the primary pathways for drug metabolism is via the CYP3A4 isoform. In humans, this pathway accounts for the metabolism of approximately 50 percent of all therapeutic drugs (Kharasch et al. 1999). Whether or not there are female/male differences in the activity of CYP3A4 is debatable. An in vitro study comparing the activity of CYP3A4 in human liver microsomes from females and males has reported that CYP3A4 activity is higher in the microsomes from females. There is in vivo evidence to suggest that systemic clearance of some drugs metabolized by CYP3A4 is faster in females than in males; however, there is also evidence suggesting that there is no difference (Kharasch et al. 1999). There are data to suggest that oral contraceptives may alter the actions of CYP3A4 and that the rate of CYP3A4-related drug clearance may differ between premenopausal and postmenopausal females (Kharasch et al. 1999). One reason for the inconsistency in CYP3A4 results may be the methods used. For example, a number of different drugs including alfentanil, diazepam, erythromycin, methylprednisolone, midazolam, prednisolone, and verapamil have been used as CYP3A4 probes. These drugs vary in their specificity for CYP3A4 and, not surprisingly, have yielded variable results. The benzodiazepine, midazolam, is reported to be one of the most reliable CYP3A4 probes available (Kharasch et al. 1999).

Kharasch et al. (1999) measured midazolam clearance on days 2, 13, and 21 of the same menstrual cycle of 11 healthy, female volunteers. The subjects ranged in age from 18 to 32 years (mean 26 years) and were within 20 percent of ideal body weight. Blood hormone levels were not measured, which is somewhat surprising as venous blood was taken for drug clearance analysis. On each experimental day, for each subject, an intravenous catheter for drug administration was positioned in a hand or arm vein, while an intravenous catheter for blood sampling was placed in a vein in the other hand or arm. The drug probe was an intravenous injection of 1 mg midazolam. Blood samples were obtained over the following 8 hours and analyzed using gas chromatography-mass spectrometry. The mean plasma concentrations of midazolam (approx. 40 ng/ml immediately following injection, decreasing to approximately 1 ng/ml after 8 hours) were not significantly different for the three test days. The volume of distribution was also unchanged. These results suggested that, for drugs metabolized primarily by CYP3A4, metabolism may be consistent across the menstrual cycle. For drugs that use other metabolic pathways, either alone or in addition to CYP3A4, menstrual cycle-related differences have been observed (Kashuba et al. 1998). For example, the antiepileptic drug phenytoin, which is metabolized by CYP2C, has been reported to be metabolized more slowly at ovulation compared with metabolism during the menstrual phase (Shavit et al. 1984). Caffeine and the related compound theophylline (used to relieve severe acute asthma) are metabolized by CYP1A2. Metabolism of both drugs has been reported to be prolonged during the luteal phase compared to the early follicular phase (Kashuba and Nafziger 1998).

Drug entry into the CNS

The transport of drugs to the brain requires that the drug be carried by the blood circulation into the brain's blood vessels, first in the arteries and finally in the capillaries. From the capillaries, the drugs must cross the capillary wall (the blood-brain barrier) to enter the brain tissue. The quantity of drug available to cross the blood-brain barrier will depend in part on the amount of blood protein and the binding of the drug molecules to protein. There is very little, if any, difference between males and females in the binding of drugs to various plasma proteins. Other blood constituents that may bind drugs include hormone binding globulins and lipoproteins and, although some sex differences have been reported, the physiological effects for most drugs would be negligible (Harris et al. 1995).

Although the blood-brain barrier has traditionally been assumed to be the same in females and males, there is some evidence to suggest that there may be sex-related differences in permeability for some drugs. Nordin (1993) reported that the CSF to plasma ratio for a metabolite of the antidepressant drug nortriptyline, but not for nortriptyline itself, differs between human females and males. The subjects in the study were 17 females (mean age 46 years) and 8 males (mean age 45 years) who were inpatients suffering from depression. CSF was collected by lumbar puncture, and blood samples were taken weekly from the second week of antidepressant treatment. When the CSF/plasma ratios of the drug were compared across subjects, the ratios were significantly higher in females compared with males. In males, but not females, the ratio was correlated with body height (which affects the clearance rate of some drugs). If both males and females had shown the same height/ratio correlation, the results could have been explained in terms of clearance alone. As the height to ratio correlation did not occur for females; however, other possible explanations, including differences in blood-brain barrier permeability, can be considered.

A study of blood-brain barrier permeability in rats using a labeled amino acid marker revealed no female/male differences in permeability in intact animals (Saija et al. 1990). However, when overiectomized (OVX) female rats were compared with rats in estrous and proestrous, the OVX rats showed significantly higher blood-brain barrier permeability. Finally, it has also been demonstrated in rats, using a fluorescent marker, that some drugs that enter the brain via the circumventricular organs (Chapter 1), rather than across the blood-brain barrier, may enter in greater quantities in females than in males (Martinez and Koda 1988).

Drug administration in pregnancy and lactation

When drugs are administered during pregnancy and lactation, the pharmacokinetics and pharmacodynamics of two bodies, not just one, have to be taken into consideration. A number of changes in a pregnant woman's physiology alter the way that she will metabolize drugs. Most of the changes are predictable given that she is supporting an extra body. Cardiac output is increased, as is kidney and liver function. The volume of both water and fat are increased, meaning that the volume of distribution is increased. Everything that happens normally happens to a greater extent during pregnancy. During lactation, the breast-feeding mother needs to consider that anything that she takes will cross into the breast milk to some extent. For example, drinking coffee during lactation can result in a wide-awake, jittery baby who gets a hefty dose of caffeine from her mother's cup of coffee.

Drug treatment of neurological disorders

Pain and analgesia

Analgesics, drugs that reduce pain, are often divided, for descriptive purposes, into prescription and nonprescription (over-the-counter) drugs. A survey of analgesic use in Sweden for 1988 and 1989 showed that females purchased more prescription and nonprescription analgesics than males. Prescription analgesics were most often purchased by people aged 45 or older, while the greatest use of nonprescription analgesics was in people aged 18 to 44. When analgesics were purchased for headache pain, both females and males purchased more nonprescription drugs. When the cause of pain was musculoskeletal, however, females were more inclined to use nonprescription analgesics than males (Antonov and Isacson 1998).

Sex differences in perception and reports of pain have been demonstrated in a number of studies. Healthy females consistently demonstrate a lower threshold for the perception of pain and also consistently report lower pain thresholds (see Craft 2007 for a review). Sex differences have also been reported in the effectiveness of different analgesics. Interestingly, the effect is not the same across drugs, suggesting very specific interactions between hormones and analgesics. The effect that estrogen has on pain is extremely complex. In some situations estrogen appears to enhance analgesia and to oppose it in other situations. The important factors for determining the effect of estrogen in any given situation seem to be the type of pain being experienced, the amount of estrogen in the system, and whether or not the estrogen level is changing or is stable (Craft 2007). It will also depend upon the type and number of estrogen receptors being activated, as the actions of the ER-α and ER-β may oppose each other. Progesterone has also been reported to modulate pain, including alleviation of diabetic neuropathy (Leonelli et al. 2007). It may be that combined estrogen and progesterone will have different effects on the experience of pain than the single hormones alone. This could be an important consideration for females taking the combined oral contraceptive pill.

In rats it has been reported that males are more sensitive than females to morphine-induced analgesia (which acts on the μ opioid receptor) (Cicero et al. 1996). In humans, nalbuphine (which acts on the κ opioid receptor) has been reported to produce better analgesia in females than in males following dental surgery (Gear et al. 1999). A randomized, double-blind study of 94 patients admitted to an emergency department in moderate to severe pain compared the responses of females (48 percent) and males (52 percent) to morphine sulfate (which acts on the μ opioid receptor, n = 46) and butorphanol (which acts on the κ opioid receptor, n = 48). At 60 minutes post-administration, the females who received butorphanol had significantly greater pain relief, as measured using a visual analog scale, than the females who received morphine. There were no significant differences between the two drugs in males (Miller and Ernst 2004).

The nonsteroidal anti-inflammatory drug ibuprofen has been reported to be more effective in reducing experimentally produced pain (electrical stimulation of the ear lobe) in males than in females (Walker and Carmody 1998).

The use of aspirin and paracetamol for mild to moderate pain relief in pregnancy is relatively safe. The only caveat with the use of aspirin is that is should be avoided in the last month of pregnancy due to its anticoagulant properties. If delivery occurred unexpectedly, then the mother could have a potential problem with bleeding due to the ingestion of aspirin.

Migraine

The class of drugs known as "triptans" is a mainstay of migraine treatment, providing effective symptomatic relief. Unfortunately, there is not yet a drug to cure migraines. The data on the use of triptans in pregnancy is limited but suggests that they are relatively safe. The metabolism of triptans may change during pregnancy, however, and the dose may need to be adjusted. The migraine treatments that are not safe are the ergotamine drugs such as ergotamine tartrate and dihydroergotamine (Soldin et al. 2008).

Neurodegeneration

In the majority of neurological disorders there is a neurodegenerative process that underlies, or occurs as a result of, the pathological condition. In Parkinson's disease, it is the loss of dopaminergic neurons in the substantia nigra; in Alzheimer's disease, it is the loss of cortical neurons. There is evidence to suggest that estrogen may act as a neuroprotectant, and, since some neurological disorders are more prevalent in older people, it is logical to ask if hormone replacement therapy (HRT) may be a useful adjunct to some drug therapies.

Because of the prevalence of neurodegenerative disorders, and the increasing average age of the world's population, increasing amounts of resources have been dedicated to studying the processes associated with neurodegeneration. There are animal models that attempt to reproduce the neurodegenerative process observed in humans. For example, there are animal models of Parkinson's disease, Alzheimer's disease, and stroke (both ischemic and hemorrhagic) that are used to study the disease process and to test possible therapeutic agents.

There are two types of degenerative processes that produce neuronal loss, apoptosis and necrosis (Figure 8.2). Apoptosis, sometimes referred to as programmed cell death, occurs in neural development when an excess of neurons is initially produced. Once developmental patterns are established, the extra neurons die off in the process of apoptosis. This process is not restricted to development, however, and occurs throughout the lifespan in response to a number of physiological triggers, including disease. During apoptosis the cell and its nucleus shrink and condense; when examined under a microscope, apoptotic cells are often characterized by dark patches of condensed cellular material called "blebbing" and fragmentation. As fragmentation occurs, the cellular debris is rapidly removed by macrophages. Necrosis, on the other hand, is the "accidental" form of cell death, often resulting from injury. When necrosis occurs, the

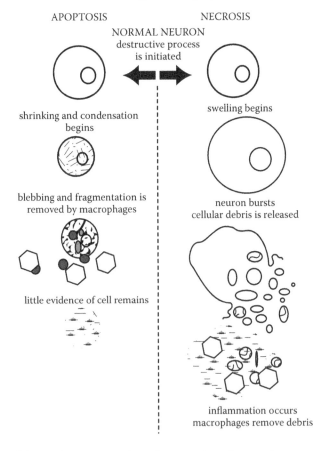

Figure 8.2 Schematic representation of apoptosis and necrosis.

cells swell and burst, spilling their intracellular material into the extracellular space, which triggers an inflammatory response.

The excitatory neurotransmitter glutamate has been associated with
cell death following stroke and other forms of neural damage. It is thought
that the axon terminals of damaged neurons release excessive amounts
of glutamate, which in turn causes damage (secondary damage) to the
neurons receiving their synaptic input (Figure 8.3). When a stroke occurs,
initially damage is restricted to the area directly around the lesion (the
core). After a period of time, however, secondary damage begins to occur
in secondary regions (the penumbra), often quite a distance away from the
area of primary damage. It is often the penumbral damage that is fatal to
the stroke victim.

The AMPA subtype of glutamate receptor (GluR1 and GluR2/3) and
estrogen receptors have been reported to be co-localized (that is, both
types of receptors are expressed on the cell surface) on neurons in several

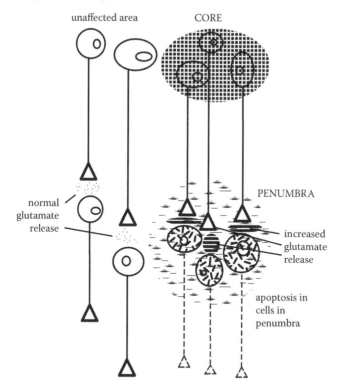

Figure 8.3 Neurons in the core region of the stroke release increased amounts
of glutamate. Neurons in the penumbral region are damaged by glutamate
excitotoxicity.

brain regions (Diano et al. 1997). Using immunohistochemical methods, neurons containing receptors for estrogen and GluR1 or GluR2/3 receptors have been identified in the amygdala and hypothalamus (Diano et al. 1997). There is also evidence to suggest that the NMDA subtype of glutamate receptor is co-localized with estrogen receptors. It has been demonstrated that mRNA for the 2D subunit of the NMDA receptor is co-localized with estrogen receptors in the hypothalamus, diencephalon, and brain stem of female rats and that the 2D subunit mRNA is up-regulated by estrogen (Watanabe et al. 1999). The effect that this co-localization of receptors may have is not well understood. However, it provides a possible mechanism for modulation of glutamatergic activity by estrogen, or estrogenergic activity by glutamate, which could be important in the development of some drug treatments. Some NMDA receptor antagonists have been demonstrated to be useful in limiting the neural damage following a stroke. It has been demonstrated that the NMDA receptor antagonist dizocilpine maleate, which at high doses produces cell death in selected neuronal populations, produces more severe damage in female rats than in male rats (Colbourne et al. 1999). This is particularly interesting, because dizocilpine maleate is one of the NMDA receptor antagonists that have been demonstrated to act as a neuroprotectant under some conditions.

In a number of experimental studies, administration of estrogen has been demonstrated to reduce the damage caused by experimentally induced stroke in rats (Dubal et al. 1998; Simkins et al. 1997; Alkayed et al. 1998). Although the protective effect has been well demonstrated, its mechanism has not been clear. It has been reported that the activity of bcl-2, a proto-oncogene that has been demonstrated to block both apoptosis and necrosis, is enhanced by estrogen treatment (Dubal et al. 1999). In this study, two groups of ovariectomized female rats were subjected to cerebral ischemia. One group received continuous estrogen treatment; the other group was treated with the oil vehicle only. Twenty-four hours after the ischemia was induced, the animals' brains were removed for analysis. In situ hybridization was used to look for changes in the expression of estrogen receptor mRNA, and RT-PCR was used to analyze the tissue for bcl-2 gene expression. Analysis of the results revealed that expression of bcl-2 was down-regulated following ischemia, but the down-regulation was prevented by administration of estrogen. In addition, expression of estrogen receptors was changed after injury.

In aging humans, estrogen has been reported to protect the hippocampus against the expected age-related reduction in hippocampal volume. An MRI study of 59 females and 38 males compared hippocampal volume between males (mean age 69 years), females taking estrogen (n = 13, mean age 67 years), and females not taking estrogen (n = 46, mean age 69 years). There were no differences in total hippocampal volumes between the females and the males. However, the right hippocampus was larger in females taking estrogen than in females not taking estrogen, and the anterior hippocampus was larger in females taking estrogen than in females not taking estrogen and the males (Eberling et al. 2003). Disturbingly, hormone replacement therapy has been associated with an increased risk of stroke that is independent of the amount of time since the therapy was initiated (Grodstein et al. 2008). This is in direct contrast to the generally held view that estrogen protects against stroke (see Brann et al. 2007 for a review).

Parkinson's disease

There is a great deal of evidence both for and against a protective or therapeutic role for estrogen in Parkinson's disease, although the consensus seems to be for a beneficial effect (Brann et al. 2007). In terms of lifetime estrogen exposure, it has been reported that females with a shorter whole life estrogen exposure due to factors such as menopause before age 36 have an higher incidence of Parkinson's disease (Rangonese et al. 2004). In animal models of Parkinson's disease, administration of estrogen has been demonstrated to protect neurons in the substantia nigra against the damaging effects of neurotoxins (Dluzen 1997; Disshon and Dluzen 1997). The question is whether or not this protective effect seen in the laboratory will translate to a useful clinical application. A study of postmenopausal female patients with Parkinson's disease of less that five years' duration suggested that there may be a clinical benefit from estrogen. The study included 138 subjects, of whom 34 subjects (mean age 61 years) had taken HRT at some stage, and 104 subjects (mean age 66 years) had never taken HRT. The average age of disease onset was significantly lower (5.6 years) in the group who had taken HRT, and the average symptom duration was significantly longer (0.5 years) in the HRT group (Saunders-Pullman et al. 1999). Another study of postmenopausal females with Parkinson's disease (n = 167) has reported that although HRT protected against the development of dementia associated with Parkinson's disease, it did not reduce the risk of developing the disease itself (Marder et al. 1998). The mainstay of treatment for Parkinson's disease is L-dopa, a precursor of dopamine, which crosses the blood-brain barrier and is metabolized to dopamine in the brain. There is little evidence on interactions between L-dopa and either estrogen or progesterone; however, it has been reported that, in rats, administration of progesterone increases L-dopa–stimulated release of dopamine from neurons in the caudate nucleus (Dluzen and Ramirez 1989).

The incidence of pregnancy in Parkinson's disease is low and the drug bromocriptine, a dopamine agonist, is rated Category B — that is, no significant risk in pregnancy. As both of these drugs are used to limit lactation, neither is indicated for use during breast-feeding. L-dopa, whether alone or in combination with a decarboxylase inhibitor, is classi-fied Category C — that is, human data are not available. It is known that L-dopa crosses the placenta and is subject to fetal metabolism, but at this time, evidence for teratogenicity has not been reported. The dopamine-releasing agent amantadine has been reported to be associated with fetal cardiovascular deformities (Rubin 2007).

Alzheimer's disease

The possible role of estrogen in the enhancement of memory and cogni-tion has received a great deal of media attention. It has been suggested that there is an "estrogen setpoint," possibly different in each person, and that when estrogen levels drop below that setpoint, cognitive function is disrupted (Arpels 1996). Not only has HRT been suggested as a way of slowing the cognitive changes associated with normal aging, it has also been suggested as a possible treatment strategy for Alzheimer's disease.

Wolf et al. (1999) tested the effects of transdermal administration of HRT on cognitive performance in a double-blind study of 38 healthy, elderly female subjects. Twenty-one subjects (mean age 70 years) received estrogen via transdermal patches placed on the back; 17 subjects (mean age 68 years) received placebo patches. Before placement of the patches and again, after two weeks of treatment, the subjects completed a range of cognitive tasks. At the same time, blood samples were taken to assess blood estrogen levels. Analysis of the results of the cognitive tests showed no differences in performance that could be attributed to estrogen treat-ment. Analysis of blood estrogen levels revealed large variations in plasma estrogen in the treated group, however. When the data for the treated group were reanalyzed with plasma estrogen levels as one of the inde-pendent variables, differences in cognitive performance became appar-ent. The performance on the verbal memory task with delayed recall was positively correlated with plasma estrogen; the higher the estrogen, the better the performance. This study is particularly interesting because the memory improvement was seen after only two weeks of treatment and the subjects, who were, on average, 17 years postmenopause, still bene-fited from estrogen.

Results from the Baltimore Longitudinal Study of Ageing indicated that HRT significantly decreases the risk of developing Alzheimer's dis-ease, as does the use of nonsteroidal anti-inflammatory drugs (Kawas et al. 1997). Of 472 peri- and postmenopausal female subjects followed for up to 16 years, 45 percent reported using HRT. There were 34 cases of Alzheimer's disease out of the 472 subjects and only 9 of the cases were HRT users. A longitudinal study of 1,124 elderly females living in New York City has also reported that HRT significantly reduces the occur-rence of Alzheimer's disease (Tang et al. 1996). Of the 1,124 subjects in the study, 167 of them (14.9 percent) developed Alzheimer's disease. Of the subjects developing the disease, 158 of them (95 percent) had never taken HRT, 5 subjects (3 percent) had taken HRT for a year or less, and 1 subject (0.5 percent) had taken HRT for more than 1 year (Figure 8.4).

Tacrine, an acetylcholinesterase inhibitor, was the original drug used for the treatment of Alzheimer's disease. Tacrine is no longer in use. It has been replaced by newer drugs that work in the same way but are better tolerated by the patients. Although acetylcholinesterase inhibitors do not ultimately alter the course of the disease, they can, in the earlier stages, improve cognitive function and enhance the quality of the patient's life. A double-blind, randomized, placebo controlled study (n = 318) on the effect of combining tacrine with HRT has reported that subjects taking the combination of drugs showed more improvement over the 30-week study than subjects taking tacrine alone (Schneider et al. 1996). A number of drugs are currently in development and in clinical trials for the treatment of Alzheimer's disease. It is hoped that from those drugs will come a cure, or at least a long-term cognitive enhancer to improve the prospects of people

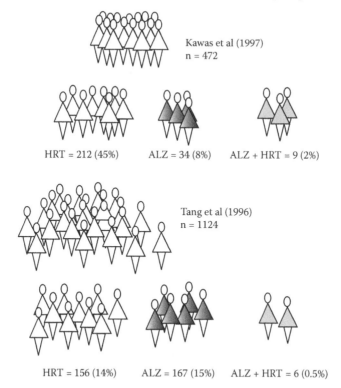

Figure 8.4 Effects of HRT on the development of Alzheimer's disease. Data from Kawas et al. (1997) and Tang et al. (1996).

with Alzheimer's disease. Not surprisingly, there is currently no literature on Alzheimer's disease treatment in pregnancy.

Multiple sclerosis

There is very little evidence for or against the use of HRT in MS. A study of experimental allergic encephalomyelitis in female rats demonstrated that administration of estradiol, but not medroxyprogesterone, inhibits the development of this experimental form of MS (Arnason and Richman 1969). The results of a questionnaire survey of 19 peri- or postmenopausal females with MS suggested that HRT may be beneficial in alleviating symptoms. The subjects responding to the questionnaire had a mean age of 56 years and a mean disease duration of 17 years. Seventy-five percent of the subjects reported an improvement in their MS symptoms with HRT (Smith and Studd 1992).

Epilepsy

The treatment of epilepsy in females requires the consideration of a number of different factors. If changes in the frequency or intensity of seizures appear to be associated with the menstrual cycle, then it may be possible to introduce hormone therapy to alter seizure susceptibility. Altering hormone levels may lead to unwanted side effects, however, such as changes in cognitive function associated with low estrogen levels. It has been suggested that the dose of antiepileptic drugs could be increased during the phases of the cycle when seizures are most likely to occur, or that other drugs, such as benzodiazepines, could be added during phases of increased seizure activity (Morrell 1999; Zahn 1999). Although all of these strategies have been reported to be useful in individual patients, none of them has been systematically tested. Two forms of hormone supplementation that may have fewer, or at least more tolerable, side effects, are the low-estrogen/high-progesterone oral contraceptive or progesterone supplements. Progesterone supplements may be particularly useful in cases where there are frequent anovulatory cycles or where the luteal phase is consistently irregular. However, progesterone supplementation is associated with cognitive side effects such as depression or sedation as well as breast tenderness and breakthrough bleeding (Zahn 1999).

There is very little evidence on the effects of menopause on epilepsy (Table 8.3). A questionnaire study of 42 postmenopausal patients (aged between 41 and 86 years) and 39 peri-menopausal patients (aged between 38 and 55) with epilepsy asked the participants about the pattern of their seizure activity and their treatment both before and after menopause. Sixteen (38 percent) of the subjects reported a premenopausal pattern consistent with catamenial epilepsy. In both groups, the patients

Table 8.3 Effects of menopause and HRT on seizure activity.

	Number of patients
Peri-menopausal patients	39
Pre-peri-menopausal catamenial epilepsy	28 (72 percent)
Effect of peri-menopause:	
Decreased seizure activity	5 (13 percent)
Increased seizure activity	25 (64 percent)
No change in seizure activity	9 (23 percent)
Patients taking HRT	8 (15 percent)
Effect of HRT	no significant effect
Postmenopausal patients	42
Pre-menopausal catamenial epilepsy	16 (38 percent)
Effect of menopause:	
Decreased seizure activity	17 (40 percent)
Increased seizure activity	13 (31 percent)
No change in seizure activity	12 (29 percent)
Patients taking HRT	16 (38 percent)
Effect of HRT	Increased seizure activity during peri-menopause

Source: From Harden et al. (1999).

with catamenial epilepsy reported that postmenopause was significantly associated with a decrease in seizure activity; however, seizure activity was reported to increase during the peri-menopause (Harden et al. 1999). There are interactions between the oral contraceptive pills and antiepileptic drugs that may lead not only to poor seizure control, but also to an unwanted pregnancy. The liver uses the same enzymes to metabolize antiepileptic drugs and oral contraceptives. Although the process of the competition for enzymes by the two classes of drugs is complex, the potential outcome is straightforward: they reduce the effectiveness of each other. There is no single answer to this problem; the effects of the competition may differ for every drug combination. The only solution is for females who have epilepsy, and want to take an oral contraceptive, to seek and follow carefully specialist advice.

An even trickier, but not impossible, combination is the use of antiepileptic drugs in pregnancy. It is well documented that prenatal exposure to antiepileptic drugs increased the risk of major birth defects from 1 to 2 percent in the normal population, to estimates of between 5 to 9 percent following antiepileptic drug exposure (Hunt et al. 2008). These effects have been well documented in the older antiepileptic drugs, such as carbamazapine and phenytoin, but there is less information available on the newer antiepileptic drugs, such as topiramate. Although the FDA

Rating Scale for Teratogenicity has been used for many years, a number of specialists in epilepsy feel that it simply does not provide adequate information on the problems encountered in treating epilepsy during pregnancy. For this reason, Epilepsy and Pregnancy Registers have been established in a number of countries in order to follow and document the course of pregnancies in epilepsy. There are two prime considerations in this respect: the damage that severe seizures could do to both mother and fetus, and the damage that the drugs could do to the fetus (Brodtkorb and Reimers 2008). One factor that changes in pregnancy is the pharmacokinetics of antiepileptic drugs. It has been reported that blood concentrations of most antiepileptic drugs decrease during pregnancy. The degree of the decrease depends upon the individual and the drug being used, but may be as much as 50 percent for some drugs. Other factors that may affect the drug level include decreased drug absorption, for example, as a result of morning sickness, increased volume of distribution, and increased metabolism (Lowe 2001). Overall, available evidence suggests that in 50 percent of cases, when drug levels are maintained, pregnancy does not change seizure frequency (see Brodtkorb and Reimers 2008, for a review). The crucial factor for a successful pregnancy seems to be careful planning and specialist advice. As some antiepileptic drugs appear to be more teratogenic than others, a drug change may be recommended, or a slow reduction in dose.

Drug treatment of psychiatric disorders

Anxiety

Until fairly recently, the treatment of anxiety has relied heavily on drugs categorized as CNS depressants (Table 8.4). Many of the drugs in this category act by potentiating the action of the inhibitory neurotransmitter GABA, and exert their effects by generally suppressing the activity of many CNS neurons. It is a characteristic of drugs in the CNS depressant category that at low doses they produce a mild sedative effect. At higher doses they cause drowsiness and sleep, until at the highest doses they can produce coma and death. Drugs in this category also have a high dependence liability (see section on drug abuse).

    CNS depressants act on the $GABA_A$ receptor, which is widely distributed on neurons throughout the brain. The $GABA_A$ receptor is particularly interesting because, in addition to a binding site for GABA, it has additional specific binding sites (allosteric binding sites) for anxiolytic drugs and hormones (Figure 8.5). Binding of ligands to the allosteric binding sites modulates the activity of the $GABA_A$ receptor that has been induced by the binding of GABA to its binding site. In humans, using PET and radioactively labeled flumazenil, it has been demonstrated that

Table 8.4 Examples of anxiolytic drugs.

Drug	Site of Action
CNS depressant	
Benzodiazepines:	
diazepam	GABA$_A$ receptor
chlordiazepoxide	
Zopiclone	GABA$_A$ receptor
Barbiturates*:	
pentobarbital	GABA$_A$ receptor
Alcohol	GABA$_A$ receptor
Atypical anxiolytic	
Buspirone	5HT-1$_A$ receptor

*Because of their serious side effects, barbiturates are no longer considered the drug of choice in the treatment of anxiety.

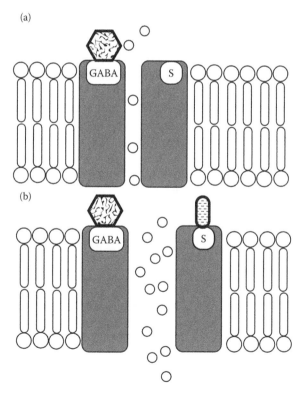

Figure 8.5 Binding of GABA to the GABA$_A$ receptor produces an influx of chloride ions (A). Binding of GABA when the steroid binding site is occupied produces an increased chloride influx (B). S, steroid binding site.

benzodiazepine binding to the $GABA_A$ receptor is reduced throughout the brains of patients with panic disorder compared with normal control subjects. The decreases were greatest in the brain regions thought to be associated with anxiety, the right orbitofrontal cortex and the right insular cortex (Malizia et al. 1998).

Steroids that act as agonists at the $GABA_A$ receptor have the effect of sedatives, inducing sleep and acting as anesthetics as well as reducing anxiety (Majewska 1992). Steroids that act as antagonists, on the other hand, may cause insomnia and anxiety, and have also been reported to improve memory (Majewska 1992).

It has been demonstrated that tetrahydroprogesterone, tetrahydrodeoxycorticosterone, and androsterone activate the $GABA_A$ receptor and that the effect of these steroids is to prolong the opening of the receptor-associated chloride ion channel (Majewska 1992) (Table 8.5) (Figure 8.5). Wilson and Biscardi (1997) have reported that the magnitude of the effects on chloride influx of tetrahydrodeoxycorticosterone, tetrahydroprogesterone, and androsterone differs between brain regions. For tetrahydrodeoxycorticosterone and tetrahydroprogesterone, the greatest effect was found in the hippocampus, with less effect in the cortex and amygdala and the least effect in the hypothalamus/preoptic area and cerebellum. For androsterone, the effect was greatest in the amygdala and hippocampus, with less effect in the cortex and hypothalamus/preoptic area and the least effect in the cerebellum. In addition, there were differences between female and male rats for tetrahydrodeoxycorticosterone in the amygdala and hypothalamus/preoptic area, where chloride influx increase was largest in males.

Two other steroids, stanozolol and 17α-methyltestosterone (synthetic anabolic-androgenic steroids), act on the $GABA_A$ receptor by inhibiting the binding of GABA agonists. Masonis and McCarthy (1995) used radioactively labeled flunitrazepam (a drug that acts as an agonist for the

Table 8.5 Effects of three steroids on $GABA_A$ receptor activity.

Steroid	Effect
Tetrahydroprogesterone	Prolongs chloride channel opening: hippocampus > cortex, amygdala > hypothalamus, cerebellum
Tetrahydrodeoxycorticosterone	Prolongs chloride channel opening: hippocampus > cortex, amygdala (m > f) > hypothalamus (m > f), cerebellum
Androsterone	Prolongs chloride channel opening: amygdala, hippocampus > cortex, hypothalamus > cerebellum

Source: From Wilson and Biscardi (1997), Majewska (1992).

benzodiazepine binding site, one of the allosteric binding sites known to modulate $GABA_A$ receptor activity) to measure the binding of the steroids in rat brain. The authors reported that although both drugs did inhibit the binding of flunitrazepam, stanozolol demonstrated much greater affinity, requiring a much lower concentration than 17α-methyltestosterone to inhibit binding. More interestingly, however, were the reported differences in binding between female and male brain tissue. By measuring the amount of radioactivity in the brain tissue and comparing changes in radioactivity across different drug concentrations, it is possible to determine some of the characteristics of the drug binding site, especially the number of different kinds of sites to which the drug is binding. In the brain tissue from male rats, the drug appeared to bind only to a single kind of binding site. In the brain tissue from female rats, however, there were two different kinds of binding sites, one with a high affinity for the drug and one with a lower affinity.

The results of binding studies of the $GABA_A$ receptor suggest a mechanism for sex differences in behaviors related to GABA function. In rats subjected to restraint stress, the pattern of binding of labeled flunitrazepam in the frontal cortex and amygdala differed between female and male rats (Farabollini et al. 1996). In the frontal cortex from female animals, there was lower affinity binding but there were more receptors, while males showed an increased affinity but no change in receptor numbers after restraint. There were no restraint-associated changes in the binding of flunitrazepam in the amygdala of either females or males. In addition, the threshold for producing seizures in rats using the GABA antagonist, bicuculline, was lower in adult male rats than in adult female rats (Bujas et al. 1997).

Because anxiety disorders are more prevalent in females than in males (Chapter 7), females also receive more prescriptions of anxiolytic drugs. As there are differences in the actions of the $GABA_A$ receptor between females and males, the possibility that anxiolytic drugs may act differently in females and males is very real. In addition, changes in drug actions across the menstrual cycle, and the possibility for interactions between oral contraceptives and drugs, have to be considered.

A randomized, double-blind study of the effects of triazolam at different phases of the menstrual cycle has yielded interesting results (de Wit and Rukstalis 1997). The participants in the study were 20 healthy, non-depressed females aged between 18 and 35 years. There were six 24-hour test sessions for each subject, two sessions in the follicular phase, two sessions at ovulation, and two sessions in the luteal phase. A test session commenced at 8:00 p.m. when the subject entered the research facility. Urine specimens were taken at that time, and the subject then stayed overnight in the facility. At 8:00 a.m. the next day, an IV catheter was placed in an arm vein and a baseline blood sample was taken to establish baseline levels for progesterone, estradiol, and allopregnanolone. The subject then completed a series of tests to assess mood, arousal, and cognitive performance. At 8:30 a.m. a capsule containing either 0.25 mg triazolam or placebo was administered. Over the following 12 hours blood samples were taken, and mood, arousal, and cognitive performance were assessed. When mood and cognitive performance were compared across the three test days, there was no clear correlation between hormone levels and test results. However, when the different phases of the menstrual cycle were analyzed independently, before drug administration in the luteal phase, there was a clear relationship between self-reported levels of arousal and the amount of allopregnanolone in the plasma. Higher levels of the hormone were associated with lower levels of arousal. In the luteal phase following triazolam administration, however, higher levels of allopregnanolone were associated with a self-reported increase in arousal. The presence of the drug reversed the relationship between the hormone and level of arousal.

An interaction between triazolam and progesterone has been reported in postmenopausal females. Kroboth and McAuley (1997) administered progesterone to subjects, followed 2.5 hours later by triazolam. In the group receiving pretreatment with progesterone, the psychomotor and sedative effects of the triazolam were significantly increased, and were independent of pharmacokinetic changes.

Although benzodiazepines as a drug class have very similar effects, there are two different metabolic pathways for benzodiazepines. Whether or not there is an interaction between benzodiazepines and another drug will depend, in part, upon whether they share the same metabolic pathway. These two processes are known as oxidation and conjugation, and oral contraceptives have been demonstrated to increase the conjugation process but inhibit oxidation. Therefore, benzodiazepines (such as alprazolam, triazolam, and diazepam) that are oxidized by liver enzymes, may have their metabolism inhibited in the presence of oral contraceptives (i.e., they may remain in the system much longer). On the other hand, the benzodiazepines metabolized via the conjugation pathway (such as lorazepam, oxazepam, and temazepam) may be metabolized much more quickly with oral contraceptives. Kroboth et al. (1997) have studied the effects of oral contraceptives on responses to lorazepam, temazepam, alprazolam, and triazolam. Clearance of alprazolam was significantly decreased in the subjects taking oral contraceptives compared with the control subjects. Clearance of temazepam was increased in the contraceptive users and clearance of lorazepam was the same as controls.

An interaction between diazepam and oral contraceptives has also been reported. An early study by Ellinwood et al. (1984) reported that females taking oral contraceptives showed significant impairment in a motor task on an "off" day compared to an "on" day. Eight female subjects aged between 21 and 26 years, and taking oral contraceptives, participated in the study. A number of tasks were used to assess cognitive and motor performance but only two tasks were discussed (Figure 8.6). The cognitive task required the subjects to learn a set of symbols, each of which corresponded to a digit between 1 and 9. During testing, the symbols were presented individually on a computer screen and the subject was required to press the keypad number corresponding to the symbol. In the motor task, a moving vertical line was presented on a computer screen. A short segment of the vertical line moved independently of the rest of the line. The task of the subjects was, using a "steering wheel" joystick, to keep the small line aligned with the large line. Subjects were tested before and after administration of 0.28 mg/kg diazepam, on cycle day 10 ("on" the oral contraceptive) and day 28 ("off" the oral contraceptive). There were no differences between day 10 and day 28 in pre-drug performance on

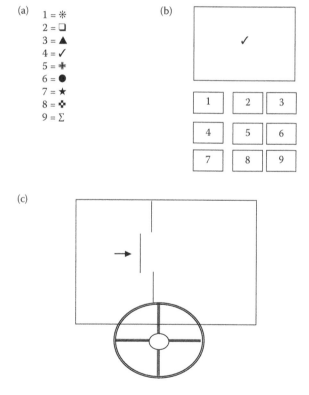

Figure 8.6 Schematic representation of tasks used by Ellinwood et al. (1984). Digit-Symbol Substitution Task: Memorize list of digits and symbols (a), when symbol is displayed, press button to select corresponding digit (b). Motor Task: Use "steering wheel" to keep moving center line aligned with fixed vertical line (c).

either task. Following drug administration, performance was significantly impaired on both tasks; however, the impairment on the tracking task was significantly greater on day 28 compared with day 10. Performance on the digit-matching task was equal on the two days. The peak impairment was also reached earlier on day 28 (20 minutes) than on day 10 (60 minutes) (Ellinwood et al. 1984).

To date, there is no evidence to suggest that the anxiolytic drug buspirone, which acts on the 5HT-1$_A$ receptor, has sex-specific effects.

The use of anxiolytic drugs in pregnancy has until fairly recently been considered relatively safe. This in itself is surprising, as they have been on the United Nations list of dangerous drugs since 1984 (Campagne 2007). In the FDA ratings, benzodiazepines are in Categories D and X. There is a good deal of evidence demonstrating adverse effects from prenatal exposure to benzodiazepines, including low birth weight, respiratory difficulties, and malformations, not to mention the benzodiazepine withdrawal syndrome. An argument for the use of these drugs has been that stress may be harmful to the pregnancy. There have been some reports that stress may be linked to miscarriage although the evidence is weak (see Campagne 2007 for a review). Apparently, the irony of the situation is lost upon those who continue to support the use of benzodiazepines in pregnancy.

Depression

There are four main categories of drugs primarily used to treat unipolar depression: tricyclic antidepressants, selective serotonin reuptake inhibitors (SSRIs), monoamine oxidase inhibitors, and atypical agents (e.g., bupropion) (Table 8.6). The first three categories of drugs act by modulating the serotonergic and noradrenergic neurotransmitter systems. While tricyclics were the first antidepressants and are still an important part of the treatment for depression, it is the SSRIs that have captured the attention of the media and the public imagination. The best-known SSRI, Prozac, has been dubbed the "happy pill" and the "personality pill" by the popular press. Drugs in the third category, the monoamine oxidase inhibitors, have regained popularity. Forms of the drug that are not specific for a single type of monoamine oxidase (type "A") interact with tyramine-containing foods with sometimes fatal consequences. The dietary restrictions required to take these drugs safely mean that they are only suitable for people who will carefully follow the prescriber's advice. Monoamine oxidase inhibitors that are specific for monoamine oxidase A do not have the severe tyramine interactions and are a very effective treatment option for some people with depression.

There is evidence demonstrating that females and males respond differently to antidepressant drugs. Given the demonstrated differences in the serotonergic system between females and males (Chapter 4), differences

Table 8.6 Examples of antidepressant drugs.

Drug	Site of Action
Unipolar Depression	
Tricyclic antidepressants	
Imiprimine	Noradrenaline and 5-HT reuptake
Desipramine	
Amitryptiline	
Selective serotonin reuptake inhibitors	
Paroxetine (Aropax)	Noradrenaline and 5-HT reuptake*
Fluoxetine (Prozac)	
Monoamine oxidase inhibitors	
Moclobemide	Inhibits monoamine oxidase type A
Atypical antidepressants	
Bupropion	Unknown
Bipolar Disorder	
Carbamazepine, valproate	Inhibition of voltage-dependent sodium conductances
Lithium	Intracellular calcium pathways

These drugs are not nearly as selective for the 5-HT transporter protein as was originally claimed (Rang et al. 1999).

in responses to drug treatments would certainly be expected. However, because females were excluded from many antidepressant drug trials, there is much less evidence than might be expected given that females receive the majority of antidepressant prescriptions. In 2000, Frackiewicz et al. conducted an extensive literature review to try and determine the extent of the data on female/male differences in antidepressant drug responses. The authors reviewed 124 published references. They found only one study that specifically evaluated female/male differences in response to antidepressants, but many studies that suggested that differences do occur (Frackiewicz et al. 2000).

A number of studies have reported pharmacokinetic differences between females and males for antidepressant drugs, particularly for tricyclics, where higher plasma levels in females have been reported. Unfortunately, many of the studies are retrospective, have very small sample sizes and have not controlled for factors such as smoking and concurrent medication usage, which can significantly alter plasma levels of a number of drugs. Overall, the evidence for sex differences in antidepressant pharmacokinetics is limited. One consistent result, however, is that pregnancy and the use of oral contraceptives may change the metabolism of some antidepressants (Yonkers et al. 1992; Frackiewicz et al. 2000).

There is evidence on female/male differences in the side effects produced by SSRIs. Sexual dysfunction is a major side effect associated with the use of SSRIs: loss of libido and delayed or failed orgasm have been reported by both males and females. In individuals suffering from chronic depression, sexual dysfunction is generally greater in females than in males. SSRI treatment in females has been reported to be associated with significantly improved sexual function, while in males it is associated with significantly worse sexual function (Piazza et al. 1997). Breast enlargement in females has been reported to occur with chronic SSRI treatment (Amsterdam et al. 1997).

A large body of anecdotal evidence, including case studies, suggests that estrogen may have antidepressant properties, particularly for menopause-associated depression. The majority of evidence to date suggests that for mild depression, estrogen may be effective; however, evidence for estrogen's effectiveness in clinical depression is lacking (Pearlstein 1995). An exception is in the case of postnatal depression, where estrogen treatment has been reported to be effective (Gregoire et al. 1996). An interesting question regarding the efficacy of HRT in the treatment of depression concerns the formulation of the HRT administered. In females with an intact uterus, estrogen needs to be supplemented by progesterone to allow the sloughing of the uterine lining and prevent the development of endometrial cancer. Progesterone acts on the $GABA_A$ receptor to increase the inhibitory influx of chloride. It has been reported to produce dysphoria in females and it has also been reported to increase monoamine oxidase activity, which could be expected to increase depression rather than alleviate it (Pigott 1999).

For the treatment of bipolar disorder, in the past, lithium was the drug of choice, and there do not appear to be sex differences in the efficacy of lithium treatment. Recently, the strategies for treating bipolar disorder have changed, and now antiepileptic drugs, such as carbamazepine and valproate, are widely used as the first choice of treatment. Lithium treatment is sometimes augmented by other drugs, such as antipsychotic drugs. When other drugs are used in the treatment of bipolar disorder, then sex differences associated with those specific drugs will also apply to the bipolar patient.

Antidepressants are not considered safe for use in pregnancy. Initial reports in 2004 warned that when administered in the last trimester, the health of the newborn could be compromised, particularly respiratory function. In 2005 and 2006, more warnings were issued. The SSRI paroxetine (now rated Category D), when taken in the first trimester, could cause serious malformations of the fetus and continuing SSRI use during pregnancy was associated with persistent pulmonary hypertension in the newborn (see Campagne 2007 for a review). The definition of Category D drugs specifies the problem of balancing the need for the drug for the mother with the risk to the pregnancy. This is certainly the case for anti-depressants. Depression is a serious, sometimes terminal disease, and treatment without drugs can be difficult. It may be the case that the use of antidepressants is necessary to allow the depressed person to become responsive enough to benefit from non-drug treatments, for example, counseling or cognitive behavioral therapy. It seems clear, however, that the use of antidepressants should be considered only after non-drug therapies have been shown to be ineffective.

Psychosis

The most compelling evidence for sex differences in response to psychoactive drugs is for antipsychotic medications. Many antipsychotic drugs act on dopamine or 5-HT receptors, and interactions between dopamine or 5-HT and estrogen are well documented (Chapter 4).

Antipsychotic drugs are classified as classical antipsychotics and atypical antipsychotics. All antipsychotic drugs act, to some extent, on the D_2 subtype of the dopamine receptor; some drugs act on other subtypes of dopamine receptors and/or other subtypes of other neurotransmitter receptors. In other words, there is no single mechanism of action guaranteed to have antipsychotic effects. Rather, it is the pattern of action on a number of different receptor subtypes that distinguishes the different antipsychotic drugs (Table 8.7).

Trying to establish an animal model of a complex human condition, such as psychosis, is fraught with difficulties, empirical and logical. How can one possibly tell if an animal is experiencing psychosis? One way of approaching the problem is to administer drugs known to produce

Table 8.7 Examples of receptor actions of some antipsychotic drugs.

Drug	Site of Action
Classical antipsychotics:	
Chlorpromazine	D_2, α-adr $> D_1$, H_1, mACh $> 5\text{-HT}_2$
Flupenthixol	5-HT_2, $D_2 > D_1$, α-adr
Haloperidol	$D_2 > D_1$, 5-HT_2, $H_1 >$ mACh, α-adr
Thioridazine	α-adr $> D_2$, mACh, $5\text{-HT}_2 > D_1$
Atypical antipsychotics:	
Clozapine	$5\text{-HT}_2 > D_1$, D_2, α-adr, H_1, mACh
Risperidone	D_2, $5\text{-HT}_2 > \alpha$-adr
Seroquel	α-adr $>$ mACh $> 5\text{-HT}_2$, D_2
Sulpiride	D_2

Source: Adapted from Rang et al. (1999, 543). α-adr, alpha subtype, adrenergic receptor; D, dopamine; H, histamine; mACh, muscarinic subtype, acetylcholine receptor; 5-HT, serotonin.

psychosis in humans to experimental animals and observe the behavior. Then, in a leap of faith, to use those same behaviors, when produced by another drug, as an indicator of the psychosis-producing potential of the new drug. Using this approach, stereotyped behavior in rodents (excessive grooming, rearing, sniffing, etc.) has come to be considered as an "animal model" of psychosis. One drug that produces the symptoms of psychosis in humans and stereotyped behavior in rats is the NMDA receptor antagonist dizocilpine maleate.

Administration of dizocilpine maleate (in mg/kg matched doses) to female and male rats has been demonstrated to produce significantly greater stereotyped behavior, and significantly greater serum and brain levels of dizocilpine maleate, in females compared with males (Andiné et al. 1999). Administration of apomorphine will also induce stereotyped behavior in rats, and haloperidol can be used to induce catalepsy. Both effects can be reduced by the administration of estradiol (Häfner et al. 1991). The escape-avoidance response is used to assess drug effects in experimental animals. In this kind of experiment, the animals are placed in an apparatus, usually a two-way shuttle box, which has distinct areas for the animals to move between. The animal is placed in one side of the apparatus and a signal (often a light) comes on followed a few seconds later by an aversive stimulus, often an electric shock. Initially, the animal leaves the chamber when the shock comes on (escape); but after training, it learns to leave when the light comes on (avoidance). Administration of antipsychotic drugs impairs the escape-avoidance response in mice, and the impairment is greater in males than in females (Monleón et al. 1998; Parra et al. 1999).

If sweeping generalizations can be made, then one sweeping generalization that it is fairly safe to make is that females are more sensitive to antipsychotic drugs than males. Females respond "better" to the drugs, but unfortunately they also experience more severe side effects, especially tardive dyskinesia (Pollock 1997; Jeste et al. 1996; Yonkers et al. 1992). Tardive dyskinesia is a syndrome of motor symptoms, especially grimacing and lip smacking that develops as a result of antipsychotic administration. Initially, if treatment is discontinued, the tardive dyskinesia will abate. If, however, treatment is continued, the tardive dyskinesia becomes a permanent, and severe, disability. It appears from the majority of the evidence that estrogen increases female sensitivity to antipsychotic drugs. However, it has been observed that tardive dyskinesia develops more frequently in postmenopausal females (Yonkers et al. 1992).

A study of hospital admissions of 65 females with schizophrenia to an acute psychiatric ward has reported that 62 percent of the patients were admitted during the late luteal phase or menstrual phase, when estrogen levels were lowest. This difference in admission rate between the low- and high-estrogen phases was interesting, but not statistically significant.

However, there was a significant difference between the two groups in the amount of antipsychotic medication required to control the symptoms. The low estrogen group needed a significantly lower dose of medication than the high estrogen group (Gattaz et al. 1994).

Most antipsychotic medications are in FDA Category D; the risk to the mother of not taking the medication must be weighed against the risk of the drug to the fetus. The current view is that the use of the drug is warranted and outweighs the risks. Overall, however, it is clear that for a number of reasons psychiatric illness during pregnancy is a risk factor for poor pregnancy outcome, and specialist care is essential for the welfare of both mother and child (see Schneid-Kofman et al. 2008 for a review).

The nontherapeutic use of drugs

Drugs that act on the $GABA_A$ receptor, such as benzodiazepines and alcohol, have a high dependence liability. Abrupt discontinuation of the drugs can result in a withdrawal syndrome that can include restlessness, anxiety, depression, and in extreme cases, seizures. Benzodiazepines are particularly liable to produce dependence, even after only a short period of administration, and current prescribing guidelines in many countries suggest limiting benzodiazepine prescriptions to 10 days. Although sex differences in liability to benzodiazepine tolerance have not been reported, worldwide the majority of benzodiazepine prescriptions are for females.

The use and abuse of alcohol raises a number of issues for females. Alcohol metabolism is via the liver enzyme alcohol dehydrogenase. The amount of alcohol dehydrogenase available is limited by its rate of synthesis. If the amount of alcohol consumed exceeds the metabolic capacity of available enzyme, the enzyme stores will be depleted and the leftover alcohol will simply stay in the bloodstream until more enzyme is synthesized. This type of elimination, zero-order kinetics, is in contrast to the first-order kinetics that applies to most drugs. With first-order kinetics, the rate of elimination is proportional to the plasma level of the drug (Figure 8.7). In practical terms, excessive alcohol consumption the night before can lead to blood-alcohol levels in the morning that are higher than the legal limit for driving a car. On a drink-per-drink basis, females reach higher blood-alcohol concentrations, and eliminate alcohol faster. The higher blood-alcohol concentrations may be due to smaller body size and/or differences in alcohol dehydrogenase activity between females and males (Mumenthaler et al. 1999). The basis for the sex differences in the rate of elimination is unclear. It has been reported that neither peak plasma concentration nor elimination of alcohol changes across the menstrual cycle, either in macaque monkeys (Green et al. 1999) or in humans (Mumenthaler et al. 1999). Excessive and prolonged alcohol consumption causes a number of health problems. In the United States it is estimated that

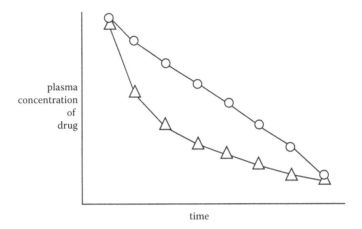

Figure 8.7 Illustration of zero-order and first-order kinetics. Open circles represent zero-order; rate of clearance of drug is independent of plasma drug concentration. Open triangles represent first-order kinetics; rate of elimination is proportional to plasma drug concentration.

24 percent of males and 5 percent of females suffer from alcohol-related disorders (Seeman 1997). Although the prevalence of the disorders is significantly lower in females (due to the lower rate of alcoholism in females), the time course of alcohol-related pathology is accelerated (Seeman 1997). Females generally start drinking later but develop the problems earlier. In addition to the alcohol-related health problems common to males and females, there is an increased risk of breast cancer associated with heavy alcohol consumption, and the adverse effects on the fetus caused by alcohol consumption during pregnancy are well publicized and well documented. For a number of years the existence of fetal alcohol syndrome was downplayed or even denied by certain individuals, and some groups, who should have known better. Now the evidence is comprehensive and difficult to ignore. Facial and palate malformations are associated with alcohol consumption during the first trimester and early in the second trimester, when the head is undergoing rapid developmental changes. Later in the pregnancy, alcohol can cause retardation and cognitive impairment, poor motor coordination, and delayed development. The development of the syndrome is not necessarily linked to the amount of alcohol consumed and even a small amount in the wrong person, at the wrong time, can have devastating results. It is odd that a number of sources still advise that a glass of wine or two is okay. It is probably safe to assume that the people making that recommendation are not the ones who have to cope with the consequences. The saddest part of fetal alcohol syndrome is that it is 100 percent preventable. The recommendation that no alcohol should be consumed in pregnancy is still the safe one.

Nicotine is the most addictive of all of the drugs of abuse. Smoking and its health-related problems were once more prevalent in males than females. Now, however, female smokers are rapidly closing the gap. In fact, according to a 1997 report, the dependence rates in the United States for alcohol and marijuana are higher in adult males, but the dependence rates for nicotine are higher in adult females (Kandel et al. 1997). A growing body of evidence suggests that smoking may have greater detrimental effects in females than in males (Cohen et al. 2007). It appears that estrogen may upregulate enzymes in the liver and lungs that convert harmless substances contained in tobacco into toxic products. The action of nicotine in the brain is as an agonist at the nicotinic subtype of the acetylcholine receptor. One interesting characteristic of the acetylcholine receptor is that it desensitizes very quickly. Repeated administration and larger doses of nicotine are required to sustain the initial response. In rats, binding studies have demonstrated that in untreated rats, or in rats withdrawn from nicotine, receptor binding is higher in females than in males. However, chronic nicotine administration results in higher receptor density in the male rat, when the measurements are taken before the drug is withdrawn (Koylu et al. 1997). It has also been reported that dopamine release in the striatum evoked by nicotine is greater in female rats than in male rats (Dluzen and Anderson 1997). In human smokers, nicotine has been reported to enhance cognitive performance relative to performance following a period of nicotine abstinence. A group of 13 female smokers were tested, following 12 hours of abstinence and following smoking, on a series of computerized tests of reaction time, motor coordination, and the Stroop test. A significant improvement in performance was seen only for the Stroop test and there was no effect of menstrual cycle phase on performance (Pomerleau et al. 1994). There appear to be sex differences in interactions between smoking and alcohol: in females nicotine decreased alcohol consumption and also decreased positive mood, while in males smoking increased alcohol consumption and raised mood (Acheson et al. 2006). Smoking during pregnancy has been reported to have an unexpected effect on the developing fetus (Paus et al. 2008). Three hundred adolescents aged between 12 and 18 years, whose mothers had smoked during pregnancy (n = 150) or not smoked (n = 150), were recruited to take part in an MRI study. One hundred and forty-five males (tobacco exposed = 66, not exposed = 79) and 155 females (tobacco exposed = 80, not exposed = 75) were tested for verbal IQ and performance IQ. There were no differences between the scores for any of the groups. The size of the corpus callosum was measured and compared between the groups. When the scans were analyzed, there was a clear decrease in the size of the corpus callosum in the tobacco-exposed females compared to the nonexposed females, but there were no differences between the two male groups.

Cannabis (marijuana and hashish) is one of the most widely used drugs in the Western world. Cannabis is obtained from the leaves and flowers of the plant *cannabis sativa*. The psychoactive component of cannabis is Δ^9-tetrahydrocannabinol. Two to twenty-two mg are required to produce psychological effects from smoked cannabis (Ambrosio 1999). The literature to date suggests that cognitive impairment is associated with a high level of cannabis use. Cannabis smoking during pregnancy has been associated with neonatal distress and sleep disturbances that can last up to three years (Dahl et al. 1995). It has also been demonstrated that cannabis is passed almost unchanged into breast milk and that if mother smokes, then baby drinks it (Riordan and Riordan 1984).

Cocaine use in males is estimated to be three times more prevalent than in females, possibly because females may experience less effect than males. Sex differences in peak plasma levels for inhaled cocaine have been reported. A study of 14 subjects (7 female, 7 male) has demonstrated that plasma levels in response to 0.9 mg/kg cocaine hydrochloride reached a significantly higher peak, and the effects were felt significantly faster, in males than in females. In females peak plasma levels were significantly higher during the follicular phase, although the subjects reported no subjective differences in euphoria between the phases (Lukas et al. 1996). A study of male and female responses to smoked cocaine reported lower subjective ratings of euphoria in females than in males and lower subjective ratings for females in the luteal phase compared with the follicular phase (Sofuoglu et al. 1999) (Figure 8.8).

Menstrual cycle–related changes have also been reported for the subjective effects of amphetamine. In a double-blind, placebo-controlled study of 16 female volunteers, aged 18 to 35, responses to oral administration of 15 mg amphetamine or a placebo were recorded for four hours following administration. Each subject was tested four times, twice in the follicular phase (amphetamine and placebo) and twice in the luteal phase (amphetamine and placebo). Blood was drawn at the beginning of each session for analysis of hormone levels. There were no baseline differences in mood between the two cycle phases. Positive feelings following amphetamine administration were reported to be significantly greater in the follicular phase, and the magnitude of the experience was related to the levels of estrogen. Interestingly, in the luteal phase, when progesterone was also present, there was less reported effect of amphetamine and the magnitude of reported response was not related to the level of estrogen (Justice and de Wit 1999). An interesting paradoxical response to amphetamine has been reported in postmenopausal females, with the response to IV administration of amphetamine being dysphoria rather than euphoria (Halbreich et al. 1981).

The potential effect of cocaine use in pregnancy has become more of a political issue than a medical one. The data on the effects of prenatal

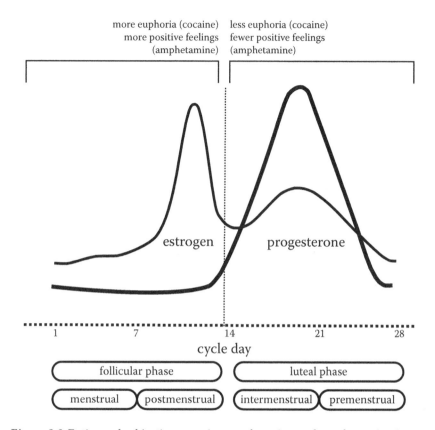

more euphoria (cocaine) less euphoria (cocaine)
more positive feelings fewer positive feelings
(amphetamine) (amphetamine)

Figure 8.8 Ratings of subjective experiences of cocaine and amphetamine in two phases of the menstrual cycle. From Sofuoglu et al. (1999) Justice and de Wit (1999).

cocaine exposure are sketchy and difficult to interpret. Many of the studies are confounded because the pregnant female was using a number of drugs in addition to cocaine, and the fetus was receiving a cocktail (see Chavkin 2001 for a review). At least some authors conclude that the effects of cocaine in pregnancy are probably not as great as those of alcohol and tobacco.

Conclusions

The following summarizes the main points to be drawn from Chapter 8:

1. Estrogen appears to have differing effects in terms of neuroprotective properties in stroke, although the evidence for a protective effect in Alzheimer's disease, and to a lesser extent in Parkinson's disease, is clearer.
2. There is limited, but sufficient, evidence to suggest that there are differences between females and males in the effects of anxiolytic,

antidepressant, and antipsychotic drugs due to differences in drug actions in the CNS. There is also evidence to suggest that the use of these drugs in pregnancy must be weighed carefully against the welfare of the mother. The majority of these drugs have been reported to have at least some degree of teratogenicity.

3. There is a female vulnerability/disadvantage for the long-term effects of antipsychotic drugs. Antipsychotic medications are generally more effective in females than in males, but females are also more likely to develop tardive dyskinesia.

4. There may be a place for estrogen supplementation of some drug therapies, for example, in Alzheimer's disease and depression.

5. Hormonal status should be a consideration in the prescription of psychoactive drugs to female patients.

Bibliography and recommended readings

Acheson, A., S. V. Mahler, H. Chi, and H. de Wit. 2006. Differential effects of nicotine on alcohol consumption in men and women. *Psychopharmacology* 186: 54–63.

Alkayed, N.J., I. Harukuni, A. S. Kimes, E. D. London, R. J. Traystman, and P. D. Hurn. 1998. Gender-linked brain injury in experimental stroke. *Stroke* 29: 159–66.

Ambrosio, E., S. Martin, C. Garcia-Lecumberri, and J. A. Crespo. 1999. The neurobiology of cannabinoid dependence: Sex differences and potential interactions between cannabinoid and opioid systems. *Life Sciences* 65: 687–94.

Amsterdam, J. D., F. Garcia-Espana, D. Goodman, M. Hooper, and M. Hornig-Rohan. 1997. Breast enlargement during chronic antidepressant therapy. *Journal of Affective Disorders* 46: 151–56.

Andiné, P., N. Widermark, R. Axelsson, G. Nyberg, U. Olofsson, E. Mårtensson, and M. Sandberg. 1999. Characterization of MK-801-induced behavior as a putative rat model of psychosis. *Journal of Pharmacology and Experimental Therapeutics* 290: 1393–1408.

Antonov, K. I., and D. G. Isacson. 1998. Prescription and non-prescription analgesic use in Sweden. *Annals of Pharmacotherapy* 32: 485–94.

Arnason, B.G., and D. P. Richman. 1969. Effect of oral contraceptives on experimental demyelinating disease. *Archives of Neurology* 21: 103–08.

Arpels, J. C. 1996. The female brain hypoestrogenic continuum from the premenstrual syndrome to menopause: A hypothesis and review of supporting data. *Journal of Reproductive Medicine* 41: 633–39.

Brann, D. W., K. Dhandapani, C. Wakade, V. B. Mahesh, and M. K. Khana. 2007. Neurotrophic and neuroprotective actions of estrogen: Basic mechanisms and clinical implications. *Steroids* 72: 381–405.

Brodtkorb, E., and A. Reimers. 2008. Seizure control and pharmacokinetics of antiepileptic drugs in pregnant women with epilepsy. *Seizure* 17: 160–65.

Bujas, M., D. Pericic, and M. Jazvinscak. 1997. Influence of gender and gonadectomy on bicuculline-induced convulsions and on $GABA_A$ receptors. *Brain Research Bulletin* 43: 411–16.

Campagne, D. M. 2007. Fact: Antidepressants and anxiolytics are not safe during pregnancy. *European Journal of Obstetrics & Gynecology and Reproductive Biology* 135: 145–48.

Cicero, T. J., B. Nock, and E. R. Meyer. 1996. Gender related differences in the antinociceptive properties of morphine. *Journal of Pharmacology and Experimental Therapeutics* 279: 767–73.

Cohen, S. B-Z., P. D. Pare, S.F.P. Man, and D. D. Sin. 2007. The growing burden of chronic obstructive pulmonary disease and lung cancer in women. *American Journal of Respiratory Critical Care Medicine* 176: 113–20.

Colbourne, F., D. Rakic, and R. N. Auer. 1999. The effects of temperature and scopolamine on N-methyl-D-aspartate antagonist-induced neuronal necrosis in the rat. *Neuroscience* 90: 87–94.

Craft, R. M. 2007. Modulation of pain by estrogens. *Pain* 132: S3–S12.

Dahl, R. E., M. S. Scher, D. E. Williamson, N. Robles, and N. Day. 1995. A longitudinal study of prenatal marijuana use. Effects on sleep and arousal at age 3 years. *Archives of Pediatric and Adolescent Medicine* 149: 145–50.

de Wit, H., and M. Rukstalis. 1997. Acute effects of triazolam in women: Relationships with progesterone, estradiol and allopregnanolone. *Psychopharmacology* 130: 69–78.

Diano, S., F. Naftolin, and T. L. Horvath. 1997. Gonadal steroids target AMPA glutamate receptor-containing neurons in the rat hypothalamus, septum and amygdala: A morphological and biochemical study. *Endocrinology* 138: 778–89.

Disshon, K.A., and D. E. Dluzen. 1997. Estrogen as a neuromodulator of MPTP-induced neurotoxicity: Effects upon striatal dopamine release. *Brain Research* 764: 9–16.

Dluzen, D. 1997. Estrogen decreases corpus striatal neurotoxicity in response to 6-hydroxydopamine. *Brain Research* 767: 340–44.

Dluzen, D. E., and L. I. Anderson. 1997. Estrogen differentially modulates nicotine-evoked dopamine release from the striatum of male and female rats. *Neuroscience Letters* 230: 140–42.

Dluzen, D. E., and V. D. Ramirez. 1989. Progesterone enhances L-dopa-stimulated dopamine release from the caudate nucleus of freely behaving ovariectomized-estrogen-primed rats. *Brain Research* 494: 122–28.

Dubal, D. B., M. L. Kashon, L. C. Pettigrew, J. M. Ren, S. P. Finklestein, S. W. Rau, and P. M. Wise. 1998. Estradiol protects against ischemic injury. *Journal of Cerebral Blood Flow and Metabolism* 18: 1253–58.

Dubal, D. B., P. J. Shughrue, M. E. Wilson, I. Merchenthaler, and P. M. Wise. 1999. Estradiol modulates bcl-2 in cerebral ischemia: A potential role for estrogen receptors. *Journal of Neuroscience* 19: 6385–93.

Eberling, J. L., C. Wu, M. N. Haan, D. Mungas, M. Buonocore, and W. J. Jagust. 2003. Preliminary evidence that estrogen protects against age-related hippocampal atrophy. *Neurobioloby of Aging* 24: 725–32.

Ellinwood, E. H., M. E. Easler, M. Linnoila, D. W. Molter, D. G. Heatherly, and T. D. Bjornsson. 1984. The effects of oral contraceptives and diazepam-induced psychomotor impairment. *Clinical Pharmacology and Therapeutics* 35: 360–66.

Farabollini, F., E. Fluck, M. E. Albonetti, and S. E. File. 1996. Sex differences in bensodiazepine binding in the frontal cortex and amygdala of the rat 24 hours after restraint stress. *Neuroscience Letters* 218: 177–80.

Frackiewicz, E. J., J. J. Sramek, and N. R. Cutler. 2000. Gender differences in depression and antidepressant pharmacokinetics and adverse events. *Annals of Pharmacotherapy* 34: 80–88.

Gattaz, W. F., P. Vogel, A. Riecher-Rössler, and G. Soddu. 1994. Influence of the menstrual cycle phase on the therapeutic response in schizophrenia. *Biological Psychiatry* 36: 137–39.

Gear, R. W., C. Miaskowski, N. C. Gordon, S. M. Paul, P. H. Heller, and J. D. Levine. 1999. The kappa opioid nalbuphine produces gender- and dose-dependent analgesia and antianalgesia in patients with postoperative pain. *Pain* 83: 339–45.

Green, K. L., K. T. Szeliga, C. A. Bowen, M. A. Kautz, A. V. Azarov, and K. A. Grant. 1999. Comparison of ethanol metabolism in male and female cynomolgus macaques (*Macaca fascicularis*). *Alcoholism: Clinical and Experimental Research* 23: 611–16.

Gregoire, A.J.P., R. Kumar, B. Everitt, A. F. Henderson, and J.W.W. Studd. 1996. Transdermal oestrogen for treatment of severe postnatal depression. *Lancet* 347: 930–33.

Grodstein, F., J. E. Manson, M. J. Stampfer, and K. Rexrode. 2008. Postmenopausal hormone therapy and stroke. *Archives of Internal Medicine* 168: 861–66.

Harden, C. L., M. C. Pulver, L. Ravdin, and A. R. Jacobs. 1999. The effect of menopause and perimenopause on the course of epilepsy. *Epilepsia* 40: 1402–07.

Harris, R. Z., L. Z. Benet, and J. B. Schwartz. 1995. Gender effects in pharmacokinetics and pharmacodynamics. *Drugs* 50: 222–39.

Häfner, H., S. Behrens, J. De Vry, and W. F. Gattaz. 1991. An animal model for the effects of estradiol on dopamine-mediated behavior: Implications for sex differences in schizophrenia. *Psychiatry Research* 38: 125–34.

Halbreich, U., G. M. Asnis, D. Ross, and J. Endicott. 1981. Amphetamine-induced dysphoria in postmenopausal women. *British Journal of Psychiatry* 138: 470–73.

Hunt, S., A. Russell, W. H. Smithson, L. Parsons, I. Robertson, R. Waddell, B. Irwin, P. J. Morrison, J. Morrow, and J. Craig. 2008. Topiramate in pregnancy: Preliminary experience from the UK Epilepsy and Pregnancy Register. *Neurology* 71: 272–76.

Jeste, D. V., L. A. Lindamer, J. Evans, and J. P. Lacro. 1996. Relationship of ethnicity and gender to schizophrenia and pharmacology of neuroleptics. *Psychopharmacology Bulletin* 32: 243–51.

Justice, A.J.H., and H. de Wit. 1999. Acute effects of *d*-amphetamine during the follicular and luteal phases of the menstrual cycle in women. *Psychopharmacology* 145: 67–75.

Kandel, D., K. Chen, L. A. Warner, R. C. Kessler, and B. Grant. 1997. Prevalence and demographic correlates of symptoms of last year dependence on alcohol, nicotine, marijuana and cocaine in the U.S. population. *Drug & Alcohol Dependence* 44: 11–29.

Kashuba, A.D.M., and A. N. Nafziger. 1998. Physiological changes during the menstrual cycle and their effects on the pharmacokinetics and pharmacodynamics of drugs. *Clinical Pharmacokinetics* 34: 203–18.

Kawas, C., S. Resnick, A. Morrison, R. Brookmeyer, M. Corrada, A. Zonderman, C. Bacal, D. D. Lingle, and E. Metter. 1997. A prospective study of estrogen replacement therapy and the risk of developing Alzheimer's disease: The Baltimore Longitudinal Study of Aging. *Neurology* 48: 1517–21.

Kharasch, E. D., D. Mautz, T. Senn, G. Lentz, and K. Cox. 1999. Menstrual cycle variability in midazolam pharmacokinetics. *Journal of Clinical Pharmacology* 39: 275–80.

Koylu, E., S. Demirgören, E. D. London, and S. Pögün. 1997. Sex difference in up-regulation of nicotinic acetylcholine receptors in rat brain. *Life Sciences* 61: 185–90.

Kroboth, P. D., and J. W. McAuley. 1997. Progesterone: Does it affect response to drugs? *Psychopharmacology Bulletin* 33: 297–301.

Lionelli, E., R. Bianchi, G. Cavaletti, D. Caruso, D. Crippa, L. M. Garcia-Segura, et al. 2007. Progesterone and its derivatives are neuroprotective agent in experimental diabetic neropathy: A multimodal analysis. *Neuroscience* 144: 1293–1304.

Lowe, S. A. 2001. Anticonvulsants and drugs dor neurological disease. *Best Practice and Research Clinical Obstetrics and Gynaecology* 15: 863–76.

Lukas, S. E., M. Sholar, L. H. Lundahl, X. Lamas, E. Kouri, J. D. Wines, L. Kragie, and J. H. Mendelson. 1996. Sex differences in plasma cocaine levels and subjective effects after acute cocaine administration in human volunteers. *Psychopharmacology* 125: 346–54.

Majewska, M. D. 1992. Neurosteroids: Endogenous bimodal modulators of the $GABA_A$ receptor. Mechanism of action and physiological significance. *Progress in Neurobiology* 38: 379–95.

Malizia, A. L., V. J. Cunningham, C. J. Bell, P. F. Liddle, T. Jones, and D. J. Nutt. 1998. Decreased brain $GABA_A$-benzodiazepine receptor binding in panic disorder. *Archives of General Psychiatry* 55: 715–20.

Marder, K., M-X. Tang, B. Alfaro, H. Mejia, L. Cote, D. Jacobs, Y. Stern, M. Sano, and R. Mayeux. 1998. Postmenopausal estrogen use and Parkinson's disease with and without dementia. *Neurology* 50: 1141–43.

Martinez, J. L. Jr., and L. Koda. 1988. Penetration of fluorescein into the brain: A sex difference. *Brain Research* 450: 81–85.

Masonis, A.E.T., and M. P. McCarthy. 1995. Direct effects of the anabolic/androgenic steroids, stanozolol and 17a-methyltestosterone, on benzodiazepine binding to the γ-aminobutyric acid$_A$ receptor. *Neuroscience Letters* 189: 35–38.

Miller, P. L., A. A. Ernst. 2004. Sex differences in analgesia: A randomized trial of mu versus kappa opioid agonists. *Southern Medical Journal* 97: 35–41.

Monleón, S., C. Vinader-Caerols, and A. Parra. 1998. Sex differences in escape-avoidance response in mice after acute administration of raclopride, clozapine, and SCH 23390. *Pharmacology Biochemistry and Behavior* 60: 489–97.

Morrell, M. J. 1999. Epilepsy in women: The science of why it is special. *Neurology* 53 (Suppl 1): S42–48.

Mumenthaler, M. S., J. L. Taylor, R. O'Hara, H-U. Fisch, J. A. Yesavage. 1999. Effects of menstrual cycle and female sex steroids on ethanol pharmacokinetics. *Alcoholism: Clinical and Experimental Research* 23: 250–55.

Nordin, C. 1993. CSF/plasma ratio of 10-hydroxynortriptyline is influenced by sex and body height. *Psychopharmacology* 113: 222–24.

Parra, A., M. C. Arenas, S. Monleón, C. Vinader-Caerols, and V. M. Simón. 1999. Sex differences in the effects of neuroleptics on escape-avoidance behavior in mice: A review. *Pharmacology, Biochemistry and Behavior* 64: 813–20.

Paus, T., I. Nawazkhan, G. Leonard, M. Perron, G. B. Pike, A. Pitiot, L. Richer, L. S. Veillette, and Z. Paussova. 2008. Corpus callosum in adolescent offspring exposed prenatally to maternal cigarette smoke. *NeuroImage* 40: 435–41.

Pearlstein, T. B. 1995. Hormones and depression: What are the facts about premenstrual syndrome, menopause, and hormone replacement therapy? *American Journal of Obstetrics and Gynecology* 173: 646–53.

Piazza, L. A., J. C. Markowitz, J. H. Kocsis, A. C. Leon, L. Portera, N. L. Miller, and D. Adler. 1997. Sexual functioning in chronically depressed patients treated with SSRI antidepressants: A pilot study. *American Journal of Psychiatry* 154: 1757–65.

Pigott, T. A. 1999. Gender differences in the epidemiology and treatment of anxiety disorders. *Journal of Clinical Psychiatry* 60 (Suppl 18): 4–15.

Pomerleau, C. S., F. Teuscher, S. Goeters, and O. F. Pomerleau. 1994. Effects of nicotine abstinence and menstrual phase on task performance. *Addictive Behaviors* 19: 357–62.

Pollock, B. G. 1997. Gender differences in psychotropic drug metabolism. *Psychopharmacology Bulletin* 33: 235–41.

Rang, H. P., M. M. Dale, and J. M. Ritter. 1999. *Pharmacology, 4th Edition.* Edinburgh: Churchill Livingstone.

Ragonese, P., M. D'Armelio, P. Aridon, M. Gammino, A. Epifanio, L. Morgante, et al. 2004. Risk of Parkinson's disease in women: Effect of reproductive characteristics. *Neurology* 62: 2010–14.

Riordan, J., and M. Riordan. 1984. Drugs in breast milk. *American Journal of Nursing* 84: 328–32.

Rubin, S. M. 2007. Parkinson's disease in women. *Disability Monthly* 53: 206–13.

Saija, A., P. Princi, N. D'Amico, R. De Pasquale, and G. Costa. 1990. Aging and sex influence the permeability of the blood-brain barrier in the rat. *Life Sciences* 47: 2261–67.

Saunders-Pullman, R., J. Gordon-Elliott, M. Parides, S. Fahn, H. R. Saunders, and S. Bressman. 1999. The effect of estrogen replacement on early Parkinson's disease. *Neurology* 52: 1417–21.

Schneider, L. S., M. R. Farlow, V. W. Henderson, and J. M. Pogoda. 1996. Effects of estrogen replacement therapy on response to tacrine in patients with Alzheimer's disease. *Neurology* 46: 1580–84.

Schneid-Kofman, N., E. Sheiner, and A. Levy. 2008. Psychiatric illness and adverse pregnancy outcome. *International Journal of Gynecology and Obstetrics* 101: 53–56.

Seeman, M. V. 1997. Psychopathology in women and men: Focus on female hormones. *American Journal of Psychiatry* 154: 1641–47.

Shavit, G., P. Lerman, A. D. Korczyn, S. Kivity, M. Bechar, and S. Gitter. 1984. Phenytoin pharmacokinetics in catamenial epilepsy. *Neurology* 34: 959–61.

Simkins, J. W., G. Rajakumar, Y-Q. Zhang, C. E. Simpkins, D. Greenwald, C. J. Yu, N. Bodor, and A. L. Day. 1997. Estrogens may reduce mortality and ischemic damage caused by middle cerebral artery occlusion in the female rat. *Journal of Neurosurgery* 87: 724–30.

Smith, R., and J.W.W. Studd. 1992. A pilot study of the effect upon multiple sclerosis of the menopause, hormone replacement therapy and the menstrual cycle. *Journal of the Royal Society of Medicine* 85: 612–13.

Sofuoglu, M., S. Dudish-Poulsen, D. Nelson, P. R. Pentel, and D. K. Hatsukami. 1999. Sex and menstrual cycle differences in the subjective effects from smoked cocaine in humans. *Experimental & Clinical Psychopharmacology* 7: 274–83.

Soldin, O. P., J. Dahlin, and D. M. O'Mara. 2008. Triptans in pregnancy. *Therapeutic Drug Monitor* 30: 5–9.

Sweet, R. A., B. G. Pollock, B. Wright, M. Kirshner, and C. DeVane. 1995. Single and multiple dose bupropion pharmacokinetics in elderly patients with depression. *Journal of Clinical Pharmacology* 35: 876–84.

Tang, M-X., D. Jacobs, Y. Stern, K. Marder, P. Schofield, B. Gurland, H. Andrews, and R. Mayeux. 1996. Effect of oestrogen during menopause on risk and age at onset of Alzheimer's disease. *Lancet* 348: 429–32.

Walker, J. S., and J. J. Carmody. 1998. Experimental pain in healthy human subjects: Gender differences in nociception and in response to ibuprofen. *Anesthesia & Analgesia* 86: 1257–62.

Watanabe, T., S. Inoue, H. Hiroi, A. Orimo, and M. Muramatsu, M. 1999. NMDA receptor type 2D gene as target for estrogen receptor in the brain. *Molecular Brain Research* 63: 375–79.

Wilson, M. A., and R. Biscardi. 1997. Influence of gender and brain region on neurosteroid modulation of GABA responses in rats. *Life Sciences* 60: 1679–91.

Wolf, O. T., B. M. Kudielka, D. H. Hellhammer, S. Törber, B. S. McEwen, and C. Kirschbaum. 1999. Two weeks of transdermal estradiol treatment in postmenopausal elderly women and its effect on memory and mood: Verbal memory changes are associated with the treatment induced estradiol levels. *Psychoneuroendocrinology* 24: 727–41.

Yates, W. R., R. J. Cadoret, E. P. Troughton, M. Stewart, and T. S. Giunta. 1998. Effect of fetal alcohol exposure on adult symptoms of nicotine, alcohol, and drug dependence. *Alcoholism: Clinical & Experimental Research* 22: 914–20.

Yonkers, K. A., J. C. Kando, J. O. Cole, and S. Blumenthal. 1992. Gender differences in pharmacokinetics and pharmacodynamics of psychotropic medication. *American Journal of Psychiatry* 149: 587–95.

Zahn, C. 1999. Catamenial epilepsy: Clinical aspects. *Neurology* 53 (Supp. 1): S34–37.

chapter 9

Into the twenty-first century

In the last six chapters, evidence has been presented that (although lean in places) clearly supports the premise that there are differences between females and males in brain structure and function, as well as in susceptibility to neurological disorders and in the effects of neuroactive drugs. Research is progressing, albeit slowly, and the media is starting to take notice. In July 2008 *New Scientist* published an article, "Sex on the Brain," in which the author, Hannah Hoag, suggested that the magnitude of the accumulating evidence on sex differences in the brain "... is pointing towards the conclusion that there is not just one kind of human brain but two" (Hoag 2008, 28).

In any kind of science, the development of the discipline progresses in stages. Initial interesting observations lead to experimentation. If the observed event occurs naturally, can it also be replicated under experimental conditions? If it can be replicated, then what are the rules governing the replication? What is required for the event to take place and, just as important, what is *not* required? What are the *necessary* and *sufficient* conditions to produce the event? Once the conditions required for the event are established, then predictions about the event can be generated and tested. Take, for example, the development of the understanding of the N-methyl-D-aspartate (NMDA) subtype of glutamate receptor.

Researchers working on the actions of glutamate observed that the activity of glutamate receptors was not consistent in their experiments, suggesting that they were dealing with more than one kind of receptor. One type of receptor seemed to follow the usual activity pattern of a receptor that was activated by neurotransmitter alone. The other (suspected) subtype of receptor, however, seemed to follow a different set of rules. Further research demonstrated that the second receptor subtype, which could be selectively activated by a synthetic substance, NMDA, had an additional requirement for activation: the receptor would only operate when the cell membrane was already partially depolarized by excitatory inputs from other types of receptors. The binding of glutamate or aspartate (or NMDA in experimental conditions) to the NMDA receptor was necessary for the receptor activation but not sufficient. The *necessary and sufficient conditions* were glutamate or aspartate binding and membrane depolarization. Once the requirements for NMDA receptor activation had been established, then hypotheses about its activity could be generated and tested. Now a great deal is understood about the NMDA receptor. It is known to be associated with learning and memory processes. It is also associated with glutamate-induced neuronal damage in conditions such as stroke. Drugs that act as antagonists at the NMDA receptor have been used as neuroprotectants in experimental models of stroke, but have also (paradoxically) been shown to induce neural damage at high concentrations. The NMDA receptor has been modeled (model building is the final stage in understanding an area of science) and the model is now used to further test hypotheses. In terms of scientific method, the NMDA receptor has been quite a success story. The story of the female brain has the potential to be much, much more exciting and successful.

The story to date

In terms of scientific development, the investigation of the female brain is at only the very beginning of the process (probably equivalent to the stage when the researchers discovered that there seemed to be more than one type of glutamate receptor). There are a lot of disparate pieces of information to be analyzed. Some of the information is consistent and seems to form a pattern; other information seems contradictory or makes no sense at all. The first step in the development of a science (and ultimately, a model) of the female brain has to be a sorting process. The estrogen table (Table 9.1) includes subheadings for the type of estrogen effect (anatomical, physiological, psychological), the nature of the effect, and just for convenience, in which chapter the information can be found. Table 9.2 provides the same "sorted" information for progesterone.

When the actions of estrogen and progesterone are organized in this way, a pattern begins to emerge. Progesterone seems to be more straightforward, so we will discuss it first. When you read down the table, the entries consistently suggest an inhibitory, CNS-depressant activity for progesterone. Most obviously, progesterone increases the inhibitory effects of GABA. This alone could account for the reported sedative effects and dysphoria produced by progesterone. It also could account for the raised seizure threshold. This is also reflected in the increased sedative effects of triazolam and the decreased effects on the CNS of the stimulants cocaine and amphetamine. Decreased alpha frequency is consistent with decreased arousal, which in turn is consistent with poorer performance on mental rotation tasks. Increases in monoamine oxidase activity could also account for feelings of dysphoria (as a result of increased metabolism of noradrenaline and serotonin). The increased suicide attempts and unstable mood when progesterone is low suggest that the loss of inhibition may have a destabilizing effect in susceptible individuals.

Estrogen is more complicated; however, the majority of the evidence points to a general action as a CNS stimulant. Interpretation of

Table 9.1 Effects of estrogen on brain function.

	Nature of estrogen effect	Chapter
Neurotransmitters:		
GABA	Increases extracellular concentrations in some brain areas	4
Serotonin	High levels may desensitize 5-HT$_1$ receptor	4
	Decreases 5-HT$_1$ binding in cortical regions	
	Increases 5-HT$_2$ binding in cortical regions	
Dopamine	Increases receptor binding in striatum	4
	Decreases amphetamine-stimulated release	
Glutamate	Upregulates NMDA receptor expression	8
EEG	Increases alpha frequency	4
Laterality	R hemisphere advantage when low	6
Psychophysics	Mental rotation best when low	5
	Increased sensitivity to musk, rose, sucrose, tactile sensations when high	5
Neurology	MS lesions worst when changing, symptoms worst when low, HRT may reduce symptoms	7
	Lowers seizure threshold, increases seizure activity	
	Reduces rate of dementia in Parkinson's disease	
	Reduces development of Alzheimer's disease	8
	Increases perception of pain (?)	
	Reduces neural damage	
Psychiatry	Increased suicide attempts when low	7
	Increased mood changes when low	
	Decreases symptoms of schizophrenia	
Drugs	Mild antidepressant	8
	Decreases monoamine oxidase activity	
	Less antipsychotic medication needed when levels are rising	
	Greater positive feelings from cocaine and amphetamine	

the increases in extracellular concentrations of GABA is a difficult issue. GABA is an inhibitory neurotransmitter; however, the effect of increased amounts of GABA will depend upon what kind of cells and receptors it is acting upon. Increased binding of serotonin to 5-HT$_2$ receptors produces excitatory effects, and, interestingly, this is in cortical regions. Increased expression of NMDA receptors might also be expected to have an excitatory effect. Consistent with the excitatory effect is the effective lowering of the seizure threshold in experimental models of epilepsy. Estrogen

Table 9.2 Effects of progesterone on brain function.

	Nature of progesterone effect	Chapter
Neurotransmitters:		
GABA	Increases inhibition	
Dopamine	Increases L-dopa-stimulated release	8
EEG	Decreased alpha frequency	
Laterality	L hemisphere advantage when high	6
Psychophysics	Mental rotation worst when high	6
Neurology	MS symptoms worst when low	7
	Raises seizure threshold	
	Causes sedation	
Psychiatry	Decreases susceptibility to panic disorder	7
	Increased suicide attempts when low	
	Increased mood changes when low	
	Decreases symptoms of schizophrenia	
Drug interactions	Increases sedative effects of triazolam (postmenopause)	8
	Increases monoamine oxidase activity	
	Produces dysphoria	
	Reduces positive feelings from cocaine and amphetamine	

cannot be the full story, however, because in catamenial epilepsy, greatest seizure activity is associated with the time when the concentrations of both estrogen and progesterone are lowest. It is not surprising that estrogen increases EEG alpha frequency, but it is a little surprising that it does not improve mental rotation. The estrogen effect in schizophrenia is interesting. If, as has been suggested, one of the problems in psychosis is failure to attend to appropriate stimuli, then perhaps the slight arousing effect of estrogen assists attention. The effect of decreasing monoamine oxidase activity is consistent with a mild antidepressant effect, which may also explain the increased suicide attempts and mood instability when estrogen is low.

The neuroprotective effects of estrogen probably account for the effects in Parkinson's disease and Alzheimer's disease. The odd one out is multiple sclerosis. Increased lesion activity seems to be associated with changes in estrogen, independent of the direction of the change. The well-documented decrease in symptoms associated with pregnancy strongly suggests that estrogen and progesterone are important considerations, however. Probably one of the first questions to ask is, "Do oligodendrocytes have estrogen receptors?"

Pregnancy and the brain

How many hormonal states exist for the "normal" adult female brain? Initially one might think three states: cycling, occasionally pregnant, and postmenopausal. From puberty to late middle age, estrogen and progesterone cycle on a monthly basis, interrupted by pregnancies when estrogen and progesterone remain high. Then in the early 50s the cycling stops and both estrogen and progesterone more or less disappear from the story. This "three state" scenario seems to fit pretty well, but is this how women's bodies are designed to function? Perhaps not. The female body evolved to support 11 to 14 pregnancies... over a much shorter lifespan. There was no evolutionary value in surviving into menopause, and until relatively recently females didn't tend to. If you estimate puberty at around 14 years of age, followed by 8 to 10 years of pregnancy, with an additional 6 to 12 months for breastfeeding following each pregnancy, there are only small intervals in the female life when normal cycling can occur. Perhaps the "usual" state is high estrogen and high progesterone, and it is the current prolonged cycling that is abnormal. If rapid changes in estrogen levels are more important than the levels themselves (e.g., in MS) then the smoother estrogen levels in pregnancy would be advantageous.

Probably the time in the cycle that most closely resembles, albeit at much lower concentrations, the hormonal balance of pregnancy, is between cycle days 18 and 24 when both estrogen and progesterone are high. But this is only a resemblance; there are a number of other factors, hormone included, that circulate in the pregnant brain. Probably the most important point here is that although many pregnant females report that they feel as though their brains are not functioning well, the experimental evidence does not support the "feeling." This is so important: pregnancy is not an obstacle to intellectual function. That myth has to go down in flames.

Instructions for using the female brain

A complex piece of equipment usually comes with comprehensive instructions for its use. A new cell phone, for example, may be accompanied by a small book and an Internet address where you can find out more. An entire afternoon can be spent playing with the new phone, sending and receiving e-mail, taking pictures, and downloading music. At the end of that afternoon, phone and owner will be well acquainted. Unfortunately, no such instructions arrive with the female brain. The new owner grows up with it, in the most literal sense, unaware that she has it until someone tells her. In fact, she will never really be aware of it except by interpretation of her own feelings and actions. This makes understanding the guidelines for use exceedingly difficult and definitely a matter of trial and error.

In writing an instruction manual, the starting place would probably be an explanation of the life expectancy of the brain and guidelines on changes in its function that will occur with use.

Congratulations on your Female Brain, 2100 Edition. This brain has been designed uniquely for you and comes with a set of functions and attributes exclusive to you alone. The life expectancy of your brain is 93 years, with a full guarantee on all parts and functions. As explained in the following instruction set (available in handheld and memory globe versions), during the first 12 to 14 years of use, several important chemicals will be unavailable for use. Please note that this chemical absence is not a malfunction but is a normal part of the run-in protocol for your brain. These chemicals will, however, be important for the functioning of the brain for the ensuing 37 to 38 years and will become available appropriately. It should also be noted that at the time that these chemicals are added to the system, and again when they are ultimately removed, there will be general disturbances in function that, although causing some discomfort and possible confusion, are part of the normal operating standard. During the time of availability each of the two chemicals, estrogen and pro-gesterone, will be available on a individual cyclic basis, with the cycle length corresponding approximately to the cycles of the earth's moon. During these chemical cycles, depending upon which chemi-cal is most available, certain functions of the brain will be enhanced. It is important for optimal performance that the user acquaint herself with the functional advantages of these cyclic changes and plan her activities accordingly. The accompanying "Estrogen and Progesterone Function Chart" is provided in an attractive wall projection format and in a convenient, carry-anywhere handheld version. Again, we welcome you to the family of Female Brain 2100 users. Fifty percent of the population are satisfied users of this product.

The user's guide above may be far-fetched, but the ideas contained in it are not. There is no reason to ignore hormonal changes and pretend that differences in brain function don't exist. There is, however, every reason to be aware of the changes and their meaning, and to exploit them when-ever possible.

Pursuing the future

It would be nice to be able to say that female and male inequality is a thing of the past, but that would be patently untrue. The restrictions and boundaries are still there, although some of the camouflage is pretty effective. The world of politics is a good example. According to Wikipedia, in 2008 only four Western countries (Switzerland, Ireland, Finland, and Iceland) had female heads of state. In one of those countries, Switzerland, females only received the right to vote in 1971, compared to 1918 for Ireland, 1906 for Finland, and 1915 for Iceland. The Wikipedia list does not include prime ministers, but if it did, New Zealand would have to be included and in the past, England and Germany. Significantly, the most powerful country in the Western world, the United States, has never had a female head of state. In fact, during the run up to the 2008 elections, there was clearly a "shock-horror" reaction from some sections of American society. "A women candidate? Oh my. Do you think she could cope?" Following the political party nominating conventions, when a female vice-presidential candidate was confirmed, the murmurings would have been hilarious, if they hadn't been so frightening. In the fine old tradition of the British political comedy *Yes, Minister,* various groups and individuals debated the future of the candidate, but not on her merits. "Ah, but she has school-aged children. What if they get chickenpox in the middle of a crisis? Of course she is perfectly capable, and in other circumstances, if the time were right, in theory, a female vice president might work."

This and similar reactions to females in positions of power and authority removes the camouflage and thin veils from a very real and enduring prejudice against females. The extent of this prejudice is surprising in 2008 and raises the question of where it comes from. It has to be deeply ingrained in the social structure to have survived.

Karen Armstrong in *The Gospel According to Woman: Christianity's Creation of the Sex War in the West* suggests an interesting answer. Christianity provides a context of woman as evil. Eve let evil into the Garden of Eden, and sex and nudity were branded as sinful at the same time. Christ's mother had to be revered and yet she had given birth. The answer was the Immaculate Conception — no sex, no sin. Throughout the history of the Church, the male saints fought off the threat of sinful encounters with lustful female saints, hell-bent on dragging them into the mud of corruption. Often the terrified males retreated to the wilderness where no corrupting females were to be found. When the occasional female was allowed in the company of chaste males, they were always asexual in appearance and manner, and therefore relatively safe. There was no place for beauty, intelligence, or wit. In the age of the Troubadours, a new female image appeared, which, on consideration, was equally asexual. This fair maid could be worshipped from afar, poetry and song could be composed in her honor, but she could never, ever be touched. In fact, the fair maid was often someone else's wife. This female image reappeared in the Victorian era when the image of the ideal female was often emaciated, childlike, asexual, and totally dependent upon the protection and support of a male. Neither image could be sustained in the real world. Thus, the ground rules for the "battle of the sexes" were established. Females were the source of sin, devious, untrustworthy, and committed to ruining males. They had to be suppressed and controlled at all cost. This view can still be found in many aspects of Western culture.

An attractive, successful female often has to run the gamut of snide remarks and inferences, from females and males alike. ("Hmmm, we know how she got ahead.") In job-interview situations, females who are the most attractive often don't get the job, even though they have the best qualifications, for example, on the (unspoken) grounds that they will probably spend more time on their appearance than the work. Then there is the other (also unspoken) barrier, "What if she gets pregnant?" At the opposite end of the spectrum, the successful male may find it necessary to possess a trophy wife in order to establish his public image. There are obvious exceptions. Take, for example, Condoleezza Rice, the U.S. Secretary of State at the time of this writing. She is intelligent, well-educated, arguably the most powerful woman in Western politics, *and she is beautiful*. In terms of Armstrong's theory, how could this happen? It is really a matter of semantics to those who might otherwise feel threatened. She is not the head of state; she is his employee who does his bidding. In other words, she becomes a surrogate trophy wife, representing her lord and master to the world. Ridiculous as that may be, it makes a safe way out for the fearful.

Changing the outcomes

The most pressing question confronting people who are concerned with ending the prejudice against females is how to proceed. In some ways, the process now started will progress on its own. Females make up at least 50 percent of the students in most medical schools. When they graduate, they will take a new, more equitable perspective with them. The members of the "old boys clubs" are getting older and slowly disappearing from the system. In their place will come new female deans, presidents, senior partners, and research directors. Many men realize and abhor the discrimination. They will continue speaking out, arguing, voting, and refusing to be shouted down by the (diminishing) male detractors. This provides the basis for a slow, gradual social change. But is slow and gradual good enough? No, at least not to many intelligent, well-educated females who have tolerated the situation long enough. There are a number of ways to improve progress, without resorting to weapons of war.

Academics, female and male, can in their teaching and research acknowledge female-male differences. For example, a lecture to second-year medical students on cancer pain is incomplete without reference to female-male differences in responses to opiates. Student pharmacists need to clearly understand the potential interactions of sedatives and progesterone. In the social sciences it is hard to imagine that a well-written lecture on, for example, psychology of color choice would exclude the factors that influence females and males, respectively. The educators probably need to be educated. Many researchers are also academics and if awareness of potential sex differences informs research design and protocols, then the knowledge base should increase rapidly. The NIH Research Initiatives discussed in Chapter 2 serve as an excellent guide for such research. For a start, just the requirement that males and females are equally represented in, *and advantaged by*, medical research should mean that the gaps in the experimental literature start to be filled. In addition, the requirement for representation of females as researchers, will ensure that, at least in the long term, attention is paid to questions of particular female importance. Another very helpful exercise would be the reanalysis of some of the existing experimental data. Often in the literature, it is clear from the description of the methods that both female and male animals were used but they were then grouped together for analysis. It is conceivable that, stored in labs around the world, are some very important data sets on male and female differences. For future research, experimental design will be a crucial issue. In order to extract the maximum amount of information from a project, the data has to be collected at the appropriate times. Ideally, every study of differences across the menstrual cycle would use blood analysis for hormone levels to establish the exact phase of the cycle. Practically, many researchers will not have the facilities or the financial support to allow blood testing. By designing experiments around (at the very least) a four-phase cycle, however, a fair degree of experimental control can be introduced. In experiments comparing female and male responses, the importance of menstrual cycle phase for the female subjects is obvious.

In health-care delivery there needs to be a conscious effort to acknowledge and respect the differences in symptomology, treatment options, and disease outcomes. Prescribers need to consider carefully hormonal states and hormonal changes if long-term therapy is being initiated. Requests to drug companies for separate prescribing guidelines for females and males would help (and it might even stimulate research on sex differences in drug actions).

Finally and most important, every female needs to understand and appreciate her unique physiology. Cycling hormones enhance different abilities at different times. Take advantage of the differences. Self-observation is easy, effective, and absolutely free. If progesterone seems to interfere with concentration, predict when it will increase and be prepared to be a little more vigilant. If you know that your memory will be enhanced for the next week, take advantage of the improvement. Cycling hormones are not tidal waves of drugs that sweep through the brain devastating all in their path. They are normal, natural chemicals that keep our bodies functioning. It is a matter of knowing and understanding, not helplessly enduring the inevitable. There is a phenomenon in the psychology of learning known as "learned helplessness." Experimental animals are trained in a situation where no matter what they do, they cannot behave in a way to receive reward or avoid punishment. Eventually the animals quit trying and often huddle in the cage ignoring food and drink. They have learned to be helpless. The suggestion that a female may be a victim of her hormones smacks of learned helplessness. Humans have not evolved to sit in the corner of the cage. Occasionally, hormones do get out of control and create difficulties. That is a case for medical intervention, not suffering in silence.

Differences are to be celebrated. They make life rich and exciting. The differences between the female brain and the male brain are just more examples of the enriching diversity in our species. We are only at the beginning of understanding how different we may be, and the next 20 years should bring astonishing discoveries and insights. We may build colonies on Mars and resorts on the moon, but for some of us, the most exciting discoveries of all will lie within our female brains.

Bibliography and references

Armstrong, K. 1986. *The Gospel According to Woman: Christianity's Creation of the Sex War in the West.* New York: Doubleday.

Hoag, H. 2008. Sex on the brain. *New Scientist,* 19 July, 28–31.

Index

Printed and bound by CPI Group (UK) Ltd, Croydon, CR0 4YY

21/10/2024

01777107-0005